"THE SCIENCE OF ACHIEVEMENT"

DECODING POTENTIAL

PATHWAYS TO UNDERSTANDING

ROBERT J. FLOWER, PH.D.

ISBN# 0-9750501-0-X
Patent pending on numerous segments
Printer: Central Plains Book Manufacturing
Published by a grant from the Gilchrist Institute for Natural Sciences

2nd Edition/Revised August 2006

An experiential guide for those seeking inner truth, understanding and meaning of life

TABLE OF CONTENTS

INTRODUCTION

When you die, God will ask just one question:
"What did you do with what I gave you?"

The following pages embrace a totally unique and accurate system for understanding and achievement. They present this format in terms of scientific findings and varying schools of philosophy. Most importantly, the format itself has common sense applications that can help you, the reader, in your everyday life.

Since the format is based on nature, it is called "natural." Moreover, since it deals with philosophy, the term "thinking" is incorporated in the name. Finally, certain innate intelligences described throughout the history of mankind and found in nature are likewise adopted. This is the genesis of Natural Thinking & Intelligence (Na.T.I., or NATI).

Throughout the book you will find various scientific theorems and postulates that lead to the world of quantum reality, where I discovered a matrix of nature's principles. In turn, this led me to varying exciting mechanisms of understanding and achievement, such as the Potential Intelligence Model, which you will read about.

Through over twenty-five years of research and investigation, I discovered that the quantum world is different from our material world. It is not that the material world is "wrong"; it's more that the daily reality we experience and understand cannot comprehend the quantum world and the way it impacts our perceived existence. An example of this is the boundless potential of nature and human imagination. Mankind still has not grasped adequately the notion of unlimited potential.

Even business recognizes this potential of human nature. Bill O'Brien, CEO of Hanover Insurance, once told the author Peter Senge (Fifth Discipline) that organizations more involved with human nature are advanced because "it has to do with the evolution of consciousness. Mankind's nature is to ascend to greater awareness of our place in the natural order. Yet, everywhere we look we see society in a terrible mess of self-centeredness, greed and nearsightedness."

O'Brien makes two key points here: greater awareness and our place in the natural order. This book will hopefully do just that–bring humankind to a greater awareness of its potential and a definite understanding of nature and its fundamental principles.

Subjective Reality vs. Objective Physics

We have known for a long time that the "mess" of humankind is a result of human frailties. Why it happens is simple: most people are too weak and totally unwilling to grow, develop, and become stronger. That's a personal choice and, thank God, not one we all embrace. For the rest of us it's a question of how to prevail, not why we don't. The answer is in the compliance with nature's laws. Why? Because of nature's absolute quality. It is impeccable.

When Albert Einstein discovered General Relativity, the cry was that "he has done away with absolutes." Einstein, however, believed the universe was not simply relative. He believed in a real, objective universe that existed independent of any observers. But if each of us sees a different universe, how is it possible to describe that universe as objective? Einstein's answer was that objectivity lay in the laws of nature. No matter how observers move, it is in accord with the same laws of nature. This very point, this sameness or consistency, is universally present not only in nature, but in those principles in nature that we have incorporated in the NATI structure. Different observers may experience different phenomena, yet the laws of nature they utilize are identical. The laws of nature are invariant according to relativity. This factor, along with theories such as the Uncertainty Principle and Probability Theory, are all found in NATI.

While relativity is based on continuous fields, quantum theory reveals nature as ***discontinuous*** (i.e., it uses quantum leaps commonly known as intuition). Further, quantum theory reveals an indeterminate yet probabilistic universe with dependant (or connected, continuous) observation. How can this be? How can dualities of this sort exist, as they so often do in our universe?

The answer lies in the fact that quantum systems are undivided wholes, or single systems. We will see this theory developed in accordance with the holographic principle later on. Since the views of scientists once again bring us to a point where science meets mysticism, we will need to explore Holographic Theory before examining how the human brain itself is perhaps the ultimate hologram and is responsible for the most important task of all: achieving potential.

In order for any theory to prevail, cutting edge ideas must evolve. Essentially, this means that societies and institutions need to focus upon new data rather than on conventional knowledge. We must be natural versus being normal. The distinctions between the two are many, the chief ones being that "normal" deals with opposition and concreteness, while "natural" deals with integration and meaning. NATI offers several key principles that enable this type of focus to occur. As we shall see, the perspectives, postulates, contradictions, and paradoxes of nature (such as those stated above) can be resolved by the NATI language of nature and various underlying transforming principles.

Violating Natural Law: A Sociopolitical Example

So how does all of this play out with daily reality? One example follows. In society today, there are thieves and cheaters. As a group, they have their own codes, follow their own rules, and have their own methods. They rationalize their position in various ways. They enter this category when, as a result of corruption within society, they fall out of their state of natural goodness and virtue and into a state of non-compliance and duplicity. They then invent their own standards of value to keep their class in line. What is created is a class distinction of rules and procedures. For example, this behavior is initiated when lawmakers themselves violate the laws governing a land. Accordingly, members of the society who recognize this see themselves as victims of this abuse. This allows them to use this as an excuse to take shortcuts and bypass the rules. Whether understandable or not, the problem is that an increasing number are not abiding by these rules. Order eventually breaks down, creating chaos and disconnection between various members of the society. This is exactly what is going on today. People make rules but then position themselves so that the rules do not apply to them.

Politicians are another example. They are supposed to be accountable and responsible, but there is little citizen recourse if they do not follow the rules and laws they themselves make. This sets up an eventual systems failure. Accordingly, the only recourse for an individual, group, or society is to vote them out of office.

So what's the answer? The answer is to redirect our focus and our concepts. If we attempt to utilize the development of potential as the major life purpose of individuals and their society, then the whole notion of corruption and destruction, right and wrong, good and evil takes on a completely different dimension. The rightness and wrongness of an issue converts to "What is the best way to accomplish development?" Sin or failure then becomes obvious and at the same time minimized as a focus. Corrective, impartial, objective criticism accepted impersonally is then the norm. The only thing to be recognized becomes "Is the objective of the development of potential being achieved for an individual, as well groups, institutions, etc."

This then is the core of NATI: the development of potential as our life's mission. Indeed, from a scientific standpoint, the universe itself is potential made visible. When we are working in harmony with the goal of the universe, partisanship and finger-pointing are eliminated. There is only growth and evolution of both the individual and society.

This book is for you. It is a powerful system for decoding the potential that is everywhere around us. You have only to focus your attention and believe that it is possible. If you do that, I promise that your life will be richer.

NATI is a holistic system capable of operating on several different levels at once. As an example, the distinguishing features of the NATI resources, when compared with competing business resources are:

- Its understanding of polarity
- Its non-judgmental approach
- Its sense of systems
- The significance of virtue
- Nature as a behavioral and thinking system

Moreover, it has the ability to implement science into real life, presenting a unification of philosophies into an overall purpose of life and achieving your fullest or highest potential!

The following pages will identity a new system (systems approach, actually) and its various parts, describing the discovery and profile of nature's principles, never before realized. It will then connect these principles with a theory of potential. We will then understand potential in a very different way.

From that point we will further identify the actual thirteen principles and relate them to our innate intelligences, which are universally present throughout humankind.

After that, we will examine these principles/intelligences through various scientific and philosophical doctrines. We will also apply these doctrines to the decoding of our life systems and our individual potential.

One further comment: there is in systems thinking a concept known as Emergent Properties. This concept deals with the parts of a whole that have a somewhat limited meaning in themselves, but come together to support and define a total whole, making all the parts significant because of the meaning of the whole. This book follows in that format. Much of the data presented will seem to be interesting but unconnected until a point at which the reader will say "Eureka!"

Lastly, there are terms you will read about throughout the book. For clarity sake, here are their interpretations:

- The 13 Principles — Basic principles discovered throughout the Natural Sciences
- Systems Thinking — Thinking in terms of Whole Picture or in Pattern Format. Deals with the whole and the parts.
- NATI — The 13 principles as applied to Human Innate Understanding — the basis for Natural Thinking Systems.
- The Science of Achievement — Realizing potential, solving problems and conflict by the way of developing creative, organizational and functional programs. The method of implementation utilizes factors such as the 13 principles, Polarity, Development as a life objective and Systems Thinking. While Psychology deals with overcoming or altering behavioral problems, achievement accomplishes results by integrating weaknesses, opposition and strength.

PART ONE

OBJECTIVE: FINDING NATURE'S PRINCIPLES

Chapter One

Nature's Paradigm

Open sesame!
—popular saying based on
A Thousand and One Arabian Nights

In order to establish a valid tool for human comprehension and personal development, certain criteria should be present. The closer the criteria are to the structure of nature, the more valid and effective the paradigm.* The criteria of this proposition was established by this researcher's findings in the works of St. Augustine, Thomas Aquinas, Leonhard Euler, Albert Einstein, Edward Reinmann and many others. These findings identified those criteria as the principles inherent in nature. As you will see, they can also be applied to human function and intelligence.

In the sense that nature contains these characteristics, it is logical that General Systems Theory, a method of utilizing biological sciences in conjunction with human behavior, would be adapted as the desired cosmological model. From this respect, this researcher has conceived the Natural Thinking & Intelligence (NATI) system.

In order to develop the theoretical basis and constructs for this model, we proceed with an overview of history relative to science, philosophy, and, in particular, quantum physics. It is important to understand the background of Natural Thinking & Intelligence systems and its cosmological basis in order to comprehend its value and process, as well as how it is in accord with

*(Langham, 1967) (Fuller, 1975)

Itzhak Bentov, David Bohm, Fritjof Capra, Karl Pribram, Rupert Sheldrake, Ken Wilbur, Gary Zukav, and other great thinkers.

The Science and Philosophy of Nature

While Plato, in the Pythagorean tradition, drew his concept from geometry, Aristotle formulated his concept of processes in nature mainly on the basis of the ways in which living organisms such as plants and animals function. We know the processes and courses of life from everyday experience. What is more obvious than to compare and explain the rest of the world, which is unknown and strange, with the familiar? According to Aristotle, the task of physics is to explain the principles and functions of nature's complexity and changes.

In modern terms, the self-organization of life was interpreted by Aristotle as a functionally governed process aiming at certain "attractors" of purposes (teleology). For instance, a tree grows out of a seed with the purpose of reaching its final form. In modern terms, the change of forms characterizing the growth of an organism is something like the (qualitative) evolution of a parameter that Aristotle called the "potentiality" of that organism. His leading paradigm of life was the idea of a self-organizing organism.

The decisive condition of modern physics was the connection of mathematics, observation, experiment, and engineering that was realized by Galileo in the Renaissance. Newton founded a new mathematical and experimental philosophy of nature that he called *Philosophiae Naturalis Principia Mathematica* (1687). Geometry and mechanics became the new paradigm of natural sciences. In the history of science, this period is called the mechanization of nature, which was imagined to be nothing more than a huge mechanical clock. The mathematician and philosopher René Descartes, together with the physicist Christian Huygens, taught that every system in nature consists of separated elements, like the cogwheels of a clock. Every effect of nature was believed to be reducible to linear causal chains like sequences of such cogwheels. Obviously, Cartesian mechanism is contrary to Aristotelian holism.

Even the physiology of life processes should be explained mechanically. The heart, for instance, was considered as a pumping machine. In general, Descartes believed that the motions of an animal and human body could be derived from the mechanism of organs "and that with the same necessity as the mechanism of a clock from the position and form of its weights and wheels."[1] The anatomy of human bodies by dissection during the Renaissance was an application of the analytical method of Descartes. According to Descartes, each system can be separated into its basic building blocks in order to explain its functions by the laws of geometry and mechanics.

The Italian physicist and physiologist Giovanni Borelli (1608-1679) founded the so-called iatrophysics as an early kind of biophysics.

Gottfried Leibniz assumed a hierarchical order of nature with a continuous scale of animation, from the smallest building blocks ("monads") to complex organisms. Leibniz tried to combine Aristotelian ideas with physical mechanics, and became one of the early pioneers for a theory of complex dynamic systems. Inspired by Leibniz, the zoologist Charles Bonnet (1720-1793) proposed a hierarchy of nature ("echelles des étres naturelles") with a measure of complexity that seems to be rather modern. Bonnet underlined "organization" as the most important feature of matter. An organization realizing the maximum number of effects with a given number of different parts is defined as the most perfect one. An organism must be described by the model of a "self-organizing being."

Following Isaac Newton's discovery of the laws of motion and gravity, a new interpretation of the universe emerged, *Determinism*. According to Determinism, the universe may be viewed as a great big clock set in motion by a divine hand at the beginning of time and left undisturbed. From its largest to its

* (Pagels, 1982)

smallest motions, the entire material creation moves in a way that can be predicted with absolute accuracy by the laws of Newton. Nothing is left to chance. The tragedy and joy of human life is already predetermined.

The famous English philosopher John Locke (1632-1704) influenced not only the epistemology and methodology of Newtonian physics, but also the political theory of modern democracy and constitution. He asked why man is willing to give up his absolute freedom in the state of nature and to subject himself to the control of political power. Locke argued that the enjoyment of the property man has in the natural state is very unsafe and insecure, because everyone else wants to take it away from him in a state of unrestricted freedom. Thus, the state of nature is unstable and will transfer the equilibrium of political forces. For Locke, the "phase transition" from the state of nature to a society with government is driven by men's intention to preserve their property.

However, by the beginning of the twentieth century a new brand of science emerged under the banner of experimental physics. What some new visionaries found was that atomic units of matter apparently behaved in random, uncontrollable ways that deterministic Newtonian physics could not account for. Aside from James Maxwell Clark and Edward Rienmann, Albert Einstein was one of the most influential in bringing about this new vision of the universe. What Einstein's discoveries reflected, in a personal sense, was his conversion from personal religion to the cosmic religion of science. He, like Spinoza, observed the universe as part of an *impersonal* energy. Einstein saw that the universe is governed by laws that can be known by us but are independent of our thoughts and feelings. The existence of this cosmic code, or the laws of material reality that are confirmed by experience, is the bedrock of faith that moves the natural scientist. The scientist sees in that code the eternal structure of reality, not as imposed by man or tradition, but as written into the very substance of the universe.[2] This cosmic code is otherwise identified as innateness, that is, the inherent components of nature that we identify as the thirteen intelligences of NATI.

Pursuing natural laws is a creative game physicists, as well as philosophers, play with nature.[3] The obstacles in this venture are the limitations of experimental techniques, as well as ignorance and fear, while the goals are the physical laws and internal logic that govern the entire universe. However, one must not become lost in confusion concerning these laws. For instance, the difference between social law and physical law is the difference between "thou shalt not" and "thou can not." No one will go to jail for violating the law of thermodynamics.[4] This idea of physical laws, as absolute, beyond the changing world, is quite remarkable.

To demonstrate this graphically, think of a globe rotating about its axis. As it moves, its appearance remains the same because a globe has symmetry; that is, it rotates about its axis, which leaves its position unchanged. This illustrates the modern idea of invariance of physical law. Symmetry implies invariance. This is why physicists are always searching for symmetry. They know if they find symmetry, it implies a new invariance—something that cannot change, an absolute. However, finding the universality of physical laws is perhaps science's greatest undertaking. Accordingly, in a comprehension sense, finding the one factor that permeates humankind totally, even if it is in varying degrees, would resolve the issue of absolute physical law. We have found that factor! We have found not only the physical laws but the social humanistic as well.

Socially, consider this: the factor is "problems"! It is beyond human comprehension that any mind on this planet does not harbor problems on a regular basis. Certainly, problems may be an issue of relativity, but problems are indeed universal, absolute factors that are inescapable for humankind. This concept of problems as an absolute issue of humankind becomes key in the realm of mind mechanics and human understanding.

Only when we accept self-responsibility by contending with our problems can we be in a valid position to achieve our potential. Problems and potential then become the mainstay of our daily pursuit of development. On a physical law basis, the factor is development! This was demonstrated by the Pulitzer Prize winner Ilya Prigogine and his study of Dissipative Structures, which we will cover shortly. Basically, Prigogine found that while matter dissipates, the universe is still growing and developing. Development is a key factor in our system!

Potential: The Missing Link Connecting Nature And Mankind

"Potential" is the most important word in the entire NATI philosophy. Potential is the "possibility of reality." It is an unsatisfied, undeveloped universal force that seeks to be realized or expressed. Potential is latent and can be manifested in an infinite number of possibilities. It represents all energy, matter, and life that have existed in the past and present (and will exist in the future). Potential brings forth all forms of life and is one of the few absolute principles of existence. In this respect, it is all-encompassing since it contains all possibilities. The scientific definition of potential is "energy stored in an object or the ability of an object to do work." For instance, a rock at the top of a cliff has more potential energy than one on the ground because of the force involved in getting from the cliff to the ground, thus converting the potential energy to kinetic energy. The notion of energy stored in an object represents an oxymoron to potential as a force. It assumes that it requires an object to take an action and places greatest significance on the object than it does on the object's potential. This presents an interesting issue concerning the original (or "first cause") nature of potential. In an ontological sense, one might say that potential emanates from the force known as the Absolute. This notion, however, leads to the age-old philosophical complaint that the nature of God could not be personal, since it would also have to be equally positive and negative. The reason for this is that it contains all possibilities including negative ones, and a humanistic god is not negative.

Given the above, potential itself is actually energy just as electricity is. It exists in a dormant stage, ready to be used at any minute. In the next moment, it can be utilized at any number of levels or degrees. There are two things, however, that potential requires regardless of how it is defined. The first is the host, through which it manifests reality. The host accomplishes this through the process of focusing, or awareness. The second is a concept or image. An example is someone turning on a light bulb by throwing a switch. This totally embraces the definition set forth of potential as being something that is imagined.

Therefore, along with the factor of imaging are the principles of awareness and expression. They are each defined as follows:

- Awareness—focus, consciousness, alertness
- Belief—the concept, the image

When both are joined, we experience expression, the form that image and awareness take. This will be explained at length in the A + B = C part of this book.

It is the author's contention that reality manifests out of the static energy of potential, which is the underlying force of the cosmos. When a constant force, such as light, is introduced into the static field, a dynamic event occurs. Within this dynamic, mass and energy are simply different manifestations of the same thing. All of the mass within us and around us is a form of bound energy.**

Infinity Within A Cell

What if nature were a system with discernible principles that enabled people to make intelligent decisions and realize their full potential? What if chaos were in some way very predictable? What if there were a science of thinking and intelligence that could help mankind make the next step in its long evolutionary trek?

** (Bohm, 1981)

I believe there is. Before we get to the system, however, a few words are in order as to how I came to be consumed with the quest for an all-encompassing model of the universe, a model that would hold true for the smallest cell as well as the largest galaxy. Although I was a perennial grad student over the years, my search directed me to obtain a Ph.D. in Philosophy, specializing in organization and systems sciences. Earlier, I engaged in graduate studies in Educational Psychology at New York University and Fordham University. Despite my academic background, however, it wasn't until 1980, when I had an epiphany about intelligence and the nature of the universe, that this book became possible. Beginning in 1980, I began a worldwide search for answers to mankind's most fundamental questions. I journeyed to great religious shrines on several continents and plunged into ever-deeper studies of physics, biology, philosophy, and theology, all of which seemed to explain one or more of the various components comprising this grand scheme we call nature or existence. As a scholar, I made presentations to the United Nations on "The New Logic" and "The Search for Natural Thinking & Intelligence." The deeper I plunged into these areas of knowledge, the more I asked myself, "Who and what *are* we? Is there a plan to life, and if so, what is it? What is truth and the nature of reality?"

I began to discover some of these answers in 1984, when I met Dr. Derald Langham, a brilliant geneticist. He was a contemporary of Buckminster Fuller and was decorated by the Venezuelan government for assisting in food plant development during the Second World War. Dr. Langham contacted me after I had written an article titled "The Master Pattern: The Concept of Living Geometry and the Stock Market." The meeting would prove to be a classic case of synchronicity, for Dr. Langham had had his own very powerful epiphany at the age of five, a vision that revealed the basic geometric and mathematical models inherent in every plant cell. I was immediately drawn to Derald's research because his vision and subsequent work as a biologist embraced not only the shape of a plant cell and all of its various growth stages, but meshed with the math and geometry of ancient sites and scriptures! Indeed, it was an idea that seemed to address the information I was uncovering, specifically in the area of megalithic sites and ancient writings.

Derald called this geometrical pattern within the cell "a living geometry." (His work, in large part, dealt with cells of the sesame seed.) The model consists of three basic parts. The first is the Creative Phase, which develops by virtue of the cell's "polar pulsing." The seed's pulse is generated to one pole and then its opposite. In other words, the cell's initial north-south pulse leads to an east-west pulse. A simple way to picture this polarity is to think of the familiar X (north-south) and Y (east-west) axes on an algebraic graph. The pulse then adopts a front-to-back movement, adding a three-dimensional Z-axis to the model. This three-dimensional pulsing of the cell completes the Creative Phase.

Next, the cell enters the Organizational Phase, in which it exhibits what can best be visualized as a Star of David inscribed within a circle. The two overlapping triangles in the Star produce a total of six points. This pattern is achieved by the wave motion of cell development. It is analogous to the power of the creative pulse. The stronger the pulse, the larger the Organizational Phase.

The third and final stage of this cellular geometry is known as the Functional Phase. Imagine a neutron traveling from the sun to the earth and becoming trapped in spiral patterns between the two poles, as well as opposite points on the equator. The spiral slows and widens as it approaches the equatorial points, while it accelerates and condenses as it nears the two poles. The neutron can be pictured as touching two points on the equator, as well as both poles, while it spirals within the earth, touching a total of four points as it moves in four different directions. This spiraling motion is characteristic of the Functional Phase. It is relative to the strength of the creative pulse and the form of the organizational wave. We are left, therefore, with thirteen aspects of this cell geometry: three in the Creative Phase, six in the Organizational, and four in the Functional.

It is crucial to remember that each of these phases has the property of Polarity, incorporating the idea of an opposite for each and every aspect of the cell's geometry. Using this idea of Polarity, it is then possible to construct a geometric cube with twenty-six equal and opposite parts: six faces, representing the three Creative Phases and their polar opposites; eight corners, representing the four Functional Phases and their polar opposites; and twelve edges, representing the six Organizational Phases and their polar opposites. *The significance of this cellular, geometric model is that in keeping with General Systems Theory, we can apply the model to human behavior. This led me to the formation of NATI, and eventually to the Science of Achievement!*

Using various lengths of connecting pipes, Derald built many representations of both the individual stages and the complete model. Although it was fascinating to look at these life-size models, few people took serious interest in his concept. I myself was personally drawn to them because of my keen interest in the megalithic sites I had visited, for I felt there was a geometric consistency between these ancient monuments and Derald's model.

Over the next several years, I spent a great deal of time with Derald. I was especially interested in the idea of opposites that the model incorporated into its overall system. This was of significance to me since the eastern philosophies I had studied-Buddhism, Hinduism, and Taoism-included ideas of opposites and ways of dealing with negativity. (In Chinese philosophy, for example, *yang* represents a positive, aggressive, masculine force, while *yin* is a negative, passive, feminine force.) Derald and I also discussed why his concept was not more widely accepted or studied. My hunch was that his colleagues did not understand the wide variety of applications that the concept could embrace.

After a while, Derald asked me if I would take his work under my wing, so to speak, and find some way to use it. In order to do this, I thought his model required a language capable of describing it for practical utilization and implementation. Although Derald didn't completely agree with this assessment, he understood that my mission was different from his, even though some of our research overlapped.

One thing that was extremely obvious to everyone who studied Derald's model was that it was a novel way of interpreting data. The Creative, Organizational, and Functional aspects struck a note of universality, *for everything that can be conceived falls into one of these areas!* I realized that Derald Langham's model defined

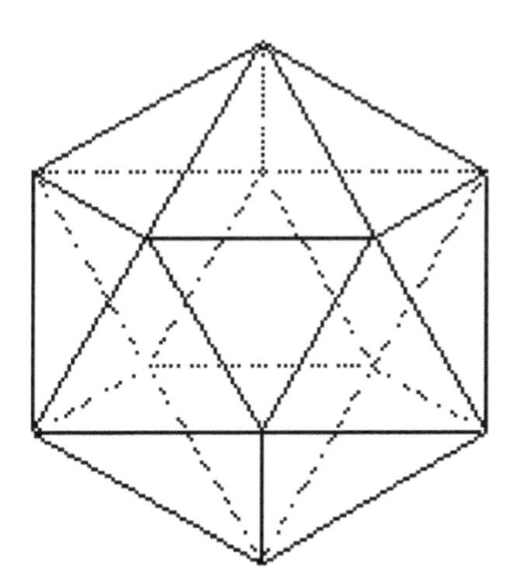

nature as an all-encompassing system capable of interpreting any kind of data that existed. *Most amazing, this incredible concept was validated by the discovery of the math and geometry of nature in various sacred scriptures and ancient megalithic sites throughout the world!*

One might think that discovering this geometrical model would be the end of a long search on my part. On the contrary, the journey was just beginning.

SUMMARY

There are three parts of the cell's Creative Phase, exhibited by the cell's pulsing. The Organizational Phase of the cell has six parts, exhibited by the cell's wave motion. The Functional Phase has four parts, exhibited by the cell's spiral motion.

The model, which describes thirteen notions in all, represents a unique way of interpreting data, for everything in existence falls into one of three categories while corresponding to cell development: Creative, Organizational, or Functional.

We now have discovered the geometry (and math) of nature found in the simple cell, thanks to Derald Langham.

Chapter Two

DEFINING INTELLIGENCE

Cogito, ergo sum: I think, therefore I am.
—René Descartes' primary self-intuition

If we are going to discuss intelligence, no matter how innate, we need to review and understand the meaning and significance of it.

If NATI is a system that uses thirteen kinds of intelligence, just exactly what is intelligence to begin with? The answer to this question will lead us farther into this revolutionary system and its conceptual foundations. (Don't be surprised as you read on if you notice that each subsequent chapter addresses multiple topics even though it may primarily focus on one issue at a time. *NATI is a holistic system capable of operating on several different levels at once.*)

The nature of intelligence has been debated for centuries. Although Greek philosophers were perhaps the first to explore the matter thoroughly, it was in the seventeenth century that two famous thinkers gave completely different answers to the question of "What is the basis for intelligence?" Rene Descartes claimed that all knowledge began with a few self-evident intuitions, the most famous being cogito, ergo sum—I think, therefore I am. John Locke, however, claimed that experience alone was the basis for any and all knowledge, saying that the human mind is a blank slate at birth, a tabula rasa. As we have seen, NATI does not favor an either/or mentality, but instead makes use of polarities, or opposites, to explain seeming contradictions. Suffice it to say at this point that there is room for both intuition and experience in any understanding of intelligence.

Let's start with some basic notions. Simply stated, intelligence is the ability to gather, recognize, and integrate data into wholes, such as concepts, and apply it competently for development. The greater a person's ability to do these things, the more successful he or she will be. What we have learned about intelligence, however, is riddled with stereotypes, and like most stereotypes, they are incorrect. We automatically assume, for example, that the nerd carrying a stack of books is intelligent, while the campus jock is inherently dumb. Neither of these assumptions, of course, is necessarily true. With Natural Thinking & Intelligence, there are categories, devoid of stereotypes, that can adequately describe every form of intelligence that, until now, has been either overlooked or categorized as something *other* than intelligence.

Currently, however, the definition of intelligence used in school systems (or for psychological testing) is terribly limited. Howard Gardner, co-director of Project Zero at the Harvard Graduate School of Education and adjunct Professor of Neurology at the Boston University School of Medicine, argues that, "the entire concept of IQ and our unitary views of intelligence have to be challenged and replaced. Our society suffers from the bias of focusing only on those human abilities that are readily testable. Merely being able to memorize and recite verses from textbooks is a narrow form of intelligence. Although it is easily measured and tested, it does not fully embody what intelligence is."[1]

Despite much study, scientists and psychologists have yet to settle on a precise characterization of intelligence. Regrettably, this hasn't dampened enthusiasm for the design and application of standardized tests purporting to measure intelligence. Given the narrow parameters we have placed on understanding intelligence, it behooves us to turn to a more established definition of the term, one found in the *Oxford Companion to Philosophy*:

> A family of intellectual traits, virtues, and abilities occurring in varying degrees and concentrations. An intelli-

> gent creature is one capable of coming up with the unexpected. An intelligent person is one in whom memory and the capacity to grasp relations and to solve problems with speed and originality are especially pronounced.[2]

As this definition implies, intelligence is not limited to the ability to store information, nor is it simply a matter of attaining verbal, social, or mathematical skills. It is also more than possessing "Superior powers of the mind," which is part of the definition given by *Webster's New World Dictionary.*[3] Intelligence, in fact, is even more than being able to use advanced skills such as analysis or calculation. More accurately, intelligence is the ability to recognize data and apply knowledge. The recognition of data, of course, implies that the data is ultimately stored, but a person's level of intelligence cannot be measured by how much information is "in storage" at any given moment; rather, the level is determined by the mind's ability to apply what is known to any given situation. It is unfortunate in the extreme that almost all students in the United States are tested to find out how much knowledge is in storage in their short-term memories on testing days. This is yet another example of problems in the educational system. Many teachers are consumed by the Great Restrictor of Fear. To avoid rocking the boat or challenging established notions about intelligence, they limit their teaching methods to the rote learning they themselves adopted in school. While critical thinking skills are mentioned in virtually every curriculum statement in the country, such skills are rarely practiced. To administrators and school boards, it is more important to finish the material outlined in lesson plans than to help students apply what they have learned in any meaningful way. Intelligence is largely approached in quantitative rather than qualitative terms, though most administrators would vehemently deny this.

Application, therefore, is crucial. Ultimately, intelligence is the act of recognizing information and then being able to satisfactorily place it into a meaningful context. To a greater or lesser degree, we all do this every day. From child to adult, from Pygmy tribe member to corporate executive, everyone expresses qualities of Natural Thinking & Intelligence. The irony is that schools suppress our natural tendencies to gain and use information in constructive ways. If information is not correctly assimilated in the acquisition stage, however, the chances increase that raw data will be misplaced in our personal schemas—our internal "big pictures"—and attempts to apply what we have learned or experienced will most likely produce confusion or undesired results. That's why it is necessary to understand the stages involved in what might be termed the evolution of raw data. In Natural Thinking & Intelligence, the best way to follow the process by which raw data is transformed into genuine understanding is through RIKU, short for Raw Data, Information, Knowledge, and Understanding. The four steps of RIKU are as follows:

- Raw data is experienced as isolated, disconnected bits of information.
- Information emerges as the bits of raw data are connected (i.e., as they form recognizable/meaningful patterns).
- Previous knowledge recognizes the connected information.
- Understanding empowers us to know how to apply the recognized information.

If all this seems too theoretical, consider the following analogy. The transformation of information into understanding is no different than the process of creating music, in which raw data consists of notes on a musical scale. By placing these notes together in various combinations, specific tones are constructed. When enough tones are juxtaposed, the result is what we call music.

Yet another way to understand RIKU is to consider the alphabet. One letter by itself doesn't have much meaning until it is connected to other letters for the purpose of forming words. In turn, the words form sentences capable of becoming paragraphs or complete stories. In both examples, isolated bits of information result in the recognition of information and, more importantly, possible ways of applying it. A decision to make a change in the way you participate in government, education, or religion would constitute using the final phase of RIKU: understanding and application. RIKU is then a prime example of emergent properties. That is, the full meaning of any bit of information is not realized until the stage at which it is understood and applied.

Being aware of the *levels* at which we acquire and retain information helps to measure how much we really understand something. Our focus can be either sharp or dull. As we have all experienced in day-to-day life, information is sometimes incorporated into our understanding in a rather shallow fashion, while at other times we process data in far greater depth. Hearing background music in an elevator is a completely different experience than consciously listening to the complex melodies and repeated motifs of a symphony. As you learn more about Natural Thinking & Intelligence, it will become clear just how important it is to have a full understanding of any piece of information in order to successfully apply it to a larger context, regardless of how slight or inconsequential the initial data may be.

When the *Titanic* sailed for New York in 1912, the danger posed by icebergs in the North Atlantic was well known, but a potent restrictor was at work on that fateful voyage: arrogance (or Ego, one of the Great Restrictors as identified by NATI). The ship was considered invincible, and the decision was made to maintain speed despite sightings of icebergs near the ship. History tells us there is every reason to believe that the tear in the ship's starboard hull might not have been as severe, (or happened at all), if information from the lookout had been processed properly by the captain.

How a small bit of data is handled can make all the difference in the world. In the case of the *Titanic*, failure to fully understand an elemental bit of data prevented what might have been a life-saving application of information. Considering the amount of data we all must process each day and the number of decisions we must subsequently make, the potential for improving our lives is truly limitless if we use a system of thinking modeled after nature itself.

To illustrate the dynamic nature of NATI (and its idea of Polarity), let us revisit the question that opened this chapter: Is intelligence a product of experience or intuitive thinking? The NATI system says that the viewpoints of both Descartes and Locke represent pieces of a larger reality. To obtain an answer that can be applied to a system of intelligence, their ideas must be synthesized. Locke and Descartes were *both* right . . . and both wrong. They each had bits and pieces of a larger picture and therefore could not apply their principles to a system of philosophy with complete understanding.

Summary

Since everyone has thirteen different kinds of intelligence, it is necessary to find a suitable definition for this key term. Far from being a question of rote learning or IQ, intelligence is the ability to recognize and apply data. Before information can truly be understood or applied, however, it must be assimilated correctly. This is done by a method called RIKU. RIKU may be summarized as follows:

- Raw data is experienced as isolated bits of information.
- Information emerges as bits of data are connected to form recognizable patterns.
- Previous knowledge recognizes the connected information.
- Understanding empowers us to apply the recognized data.

This is also identified with the nature of Emergent Properties as set forth in the Introduction and with RIKU.

Chapter Three

Building An All Inclusive, Comprehensive System

The system under consideration (NATI) is the coming together of abstract and concrete, spirit and flesh, me and you, and yes, SELF and self. It is a melding of science and the philosophical verities which together generate an absolute direction following nature's laws.

A System of Unification

In order for any system to be complete, perfect, and whole, it needs to be all-encompassing! The geometry of nature, Dr. Langham found, is classified into three general groups identified as the Creative or (Planning part), the Organizational, and the Functional. Ever since I discovered this format in 1984, others, as well as myself, have never been able to find any piece of data that could not be classified into one of these categories, and there have been several thousand that have tried!

We therefore have the basis for an all-encompassing system, a universal language so to speak! From this system we can create, plan, or organize any other system and deal with any known issue. We now have a system that enables the following: planning and creating, organizing, functioning, opposition, I.Q., kinetic intelligence, intuition, emotional intelligence, rules, steps or processes, parts, the entirety, feedback, measurement, action or performance, spirit, materialism, science, philosophy, etc.

Systems thinking is viewing things (data, experiences, situations) as a whole. This is why we place emphasis on theories such as Holography, Wholeness, and Complex Adoptive Systems.

NATI has two separate but connected aspects. It is both a science and a philosophy. This new philosophy of understanding the universe is based on the premise that while reality is unified, Polarity is an essential part of that cosmic harmony. It says that all things are interrelated. In other words, Natural Thinking & Intelligence is a system of wholeness, a system so pervasive that it has deep roots in both Eastern and Western thought. Moreover, NATI defines certain fundamental intelligences that are innate, not only to nature as a whole, but also to every human being on the planet. In a real sense, NATI is a hologram reflected in the structure of a simple plant cell. It is an absolute that nevertheless includes variability. It can, in fact, generate and explain the millions of variables comprising "chaos."

It is this holographic nature of NATI, with its inclusion of Polarity, which helps unify Eastern and Western philosophical traditions. In the West, people take great pride in the virtues of self-reliance and independence. This is reflected in their basic religious orientation as well. Westerners recognize God as being far greater than the self, a being who cares about them and judges them as individuals. The drawback to this mindset is that people in Western cultures develop mental barriers that keep them separate from the rest of the world. While this tradition of independent thinking encourages Westerners to be strong and enterprising, it also leads to serious problems, including loneliness, alienation, uncertainty, and guilt. It is probably no accident that industrialized nations in the West, placing strong emphasis on performance and assembly-line productivity, have higher rates of alcoholism, suicide, divorce, and depression.

By contrast, Eastern thought considers the individual to be part of a much larger reality. That which binds all humans together is ultimately mere consciousness, a oneness from which all individual expression emerges. The ultimate goal of Buddhism, for example, is to return to a state of pure perfection and bliss—

Nirvana—the cosmic merging of an individual's space-time consciousness with that of the eternal. Accordingly, demographic data indicates that oriental populations enjoy longer life expectancy and lower mortality rates.

NATI connects Eastern and Western mindsets quite efficiently by acknowledging the importance of the individual, but only inasmuch as he or she is part of a much larger scheme of existence. The two views are not mutually exclusive, and one might postulate that this is exactly what the Gospel of St. John meant in the words "I am in the Father and the Father is in me" (John 14:10). Later in the New Testament, this unity with the Father is extended to all of mankind when St. Paul says in his letter to the Galatians that we are brothers of Jesus, who is therefore the first of many sons. The implications of this theological principle are profound and are related to Eastern thought in more than just a passing way, for we see in the writings of St. John and St. Paul the idea that humans are part of what has become known in metaphysics as non-local (universal) intelligence. This notion was largely ignored in the Christian tradition until the twentieth century, when the French Jesuit paleontologist Pierre Teilhard de Chardin wrote of the emergence of a global consciousness, which he termed the noosphere. My research has led me to a connectivity between notions of the Father, the noosphere, Plato's World Soul, and Aristotle's Absolute. What is the connecting factor? It's potential!

Potential As A Whole And Its Parts

The idea of Collective Individuality, which is a key philosophy of NATI, refers to reality as a collection of individual experiences. That is, the collective of humanity as a whole affects each individual's experiences and vice versa. Thus, our experiences may be viewed either subjectively or objectively since the distinction between "in here" and "out there" becomes arbitrary, at best. Such polarities still exist within a holistic model, of course, but meaning becomes highly dependent on one's intent and point of observation. To put it in more colloquial terms, "Beauty is in the eye of the beholder."

A corollary to this idea of Collective Individuality is that we cannot look solely to our own feelings and impressions to give us a reliable sense of reality. In order to realize greater achievement, we paradoxically must look deeper than our individual personalities. Doing so allows us to bypass impulses that are both misguided and self-centered. It releases us from being a slave to what I call the Great Restrictors: Fear, Ego, Ignorance, and Self-deception. When we look beyond ourselves, there is a merging of individual consciousness with the universe, a process during which we can experience ourselves as interdependent parts of a whole. Such encounters, or intersections between the individual and the absolute (potential), are actually nothing new, for they are what have come to be known as mystical experiences and span both Eastern and Western traditions. The Buddha's fabled transcendence and enlightenment are remarkably similar to the experiences of medieval mystics who felt their individuality merge with a larger reality.

We can even see these moments of transcendence on a smaller scale in more contemporary settings, as when athletes are able to block out extraneous sensory input to such a degree that they feel themselves to be functioning on a totally different plane of reality. The transcendence of these experiences through the ages is why we can say that NATI is a universal language devoid of all doctrines and judgmental conclusions.

When age-old confrontations between doctrines or disciplines are eliminated—East vs. West or science vs. religion—a person's dormant spiritual energy is free to realize limitless potential. In NATI, this activated energy can then be expressed according to a principle called Directional Judgment, meaning that the momentum of an individual's thinking or action is directed toward abundance, acceptance, balance, harmony, beauty, compassion, courage, strength, honesty, truth, joy, love, cooperation, and clarity—essentially virtues defined by NATI! Furthermore, this momentum enables people to reduce stress, improve concentration, boost confidence, increase efficiency, and see issues in a larger context. Effectively, this also relates to the judging of a focused direction rather than a state of comparison.

In short, NATI, as a model of nature, enables the development of potential.

Development and potential: these are words you will see frequently as you read on. And if you are bold enough to leave prejudice and fear behind, you will begin to see a new way of thinking emerge, a new paradigm. As Marilyn Ferguson says in *The Aquarian Conspiracy*, "A new paradigm involves a principle that was present all along but unknown to us. It includes the old as a partial truth, one aspect of How Things Work, while allowing for things to work in other ways as well. By its larger perspective, it transforms traditional knowledge and the stubborn new observations, reconciling their apparent contradictions."[1]

The old and the new polarities and apparent contradictions are already reconciled in nature as seen in the geometry of the smallest cell. Is it possible that nature has been trying to tell us something for countless generations? General Systems theorists, as well as scientists studying quantum mechanics think so, and through Natural Thinking & Intelligence, we can find out what that "something" is.

Summary

The geometrical model of the cell has given birth to a system of understanding the universe. This system, called NATI, states that all things are both polarized and yet connected. The system has deep roots in both Eastern and Western philosophy, accepting the Western notion of individual identity, as well as Buddhist or Taoist notions of oneness and unity. Individuals, therefore, are important inasmuch as they are a part of a larger reality. In what NATI calls Collective Individuality, experiences are both subjective and objective. Collective individuality is the participation of the parts and the whole in a common objective. In our case, that objective is development.

When we are able to see ourselves as part of a more comprehensive, non-local (universal) intelligence, we are free to abandon the great restrictors of Fear, Ego, Ignorance, and Self-deception for the purpose of using a natural (spiritual) energy to pursue limitless potential. This activated energy is experienced through Directional Judgment, which allows people to utilize the qualities of virtue in daily life.

Directional judgment is measurement and opinions that have to do with the direction or path one is pursuing rather than condemnation, classification, or hierarchy.

We have modeled our system after the science and structure of nature and potential. Our comprehension can then be based upon absolute irrevocable factors!

Chapter 4

The Tao and Science of Potential

Lay down all thought,
Surrender to the void . . .
—John Lennon

The next step in defining our universal model is based on the consideration that if the thirteen principles are throughout nature, they may well be present in "pre-nature". This pre-nature we have identified as potential. Attempts throughout history to understand and/or implement nature have failed because of the absence and significance of potential and its development.

Wu vs. Yu

The concept of potential in any of its applications or contexts is very much future-oriented in its connotation. To understand potential further, we shall narrow our focus in this chapter to quantum physics and the *Tao te Ching*, written in China by Lao Tzu over twenty-five centuries ago. Pick up any contemporary book on quantum physics, and the chances are good that you will find references to the great spiritual teachers of India or China. Pick up any book on mysticism or Eastern thought, and the chances are overwhelming that quantum physics will be mentioned somewhere in its pages.

Ching simply means "a book," while *Tao* means "the way." *Te* implies virtue or character. The title of Lao Tzu's work, therefore, literally connotes a virtuous path through life. The book describes man's relationship to the environment, emphasizing the balance between one's inner life and the outer rhythms of nature. In advocating this synergy of inner and outer energies, the *Tao* seeks to teach men how to practice patience, cooperation, peace, health, and problem solving. Practitioners Taoism claim they are able to find new sources of joy and creativity in their lives through daily meditation, a silencing of the chaotic thoughts that result from the frenetic pace of life in the twenty-first century. Lao Tzu writes:

The Tao is mysterious, unfathomable,
Yet within is all that lives,
Unfathomable, mysterious,
Yet within is the essence.
—Tao 21

This essence is nothing less than potential, and it is this potential that meditators seek to contact as they empty their minds.

The *Tao* also expresses a fundamental principle, which is that creative potential, called *wu*, is eternal non-being. Created existence, however, manifests itself as *yu*. The paradox is that existence is both being and non-being—potential and the realization of potential. Put in other terms, reality is a great void—a nothingness—but the nothingness contains a force, called *chi*, with which all of life is imbued. The *Tao* teaches that existence is pregnant with possibility (potential energy) that must be released from the void in order to finally be expressed as reality. It is analogous to the old Newtonian law of motion, which says that every resting object contains potential energy, the energy of future motion. The energy in a ball is only released when it starts to roll. In the same way, the potential energy of *chi* is released only when we draw upon its power in our lives to create new patterns and cycles of growth. From nothingness, we literally create the world.

Whenever we come upon something that ignites a passion within us, spurring us to pursue new avenues of interest or activity, we are encountering *wu*. We are confronting the limitless potential inherent in the void. When we choose to act and actually pursue an endeavor, we are transforming potential into *yu*, the tangible world.

It is worth noting that many researchers believe that *wu* is centered in the right hemisphere of the brain.[1] While the left hemisphere processes speech, analysis, and logical thinking, the right hemisphere apprehends the world in a symbolic, intuitive fashion. This is in keeping with a fundamental teaching of the *Tao*.

The Tao that has been charted is not the eternal way.
A word we can define is not the eternal Word.
—Tao 1

Lao Tzu believed that the potential that elevated man and helped him to transcend normal thought patterns was intuitive in nature. Modern psychology confirms this in documenting how "thinking outside of the box" has resulted in so many new ideas and advances in technology.

The Quantum World

For centuries, Western civilization regarded the universe as mechanistic in nature as implied by Newton's laws of motion, which were based on cause and effect relationships manifested in solid, three-dimensional space. In other words, reality was what could be observed and scientifically measured.

This view of a clockwork universe changed drastically in 1905 when Albert Einstein submitted a paper to Max Planck, editor of *Annals of Physics*. The extremely technical paper was Einstein's Special Theory of Relativity, the basis for the famous equation $E=mc^2$.[2] The theory declared matter and energy to be interchangeable, a concept that, needless to say, could not be easily envisioned by the very grounded, rationalistic Newtonians.[3]

To understand how this radical new way of thinking of matter and energy impacts our notions of potential, we must take a whirlwind tour, a mental jaunt analogous to the European packages that show travelers ten countries in three days. Pack the bags in the left hemisphere of your brain, therefore, but beware. At the end of our trip, you will have to evaluate where you have been with your right hemisphere.

Sir Isaac Newton declared that light consisted of a stream of particles, but one of his colleagues, Dutch physicist Christian Huygens, argued that light was a wave. It was English physicist Thomas Young who proved Newton wrong in the early 1800s with his now-famous double-slit experiment.[4]

In the double-slit experiment, a beam of light is aimed at a solid barrier separated by a screen with two identical narrow openings.[5] One might expect that closing one of the slits would result in fewer rays of light reaching the screen, and that those which did would be clustered in a location that corresponded to the open slit. That's not what happens, however. With one slit open, light manages to disperse across the entire backdrop. As if this is not bizarre enough, light shows an interference pattern when both slits are open instead of politely striking the areas directly behind the slits it has allegedly passed through. It creates a wave pattern and behaves in the same way that waves at a beach behave (washing around rocks so that they are spread uniformly on the sand). The double-slit experiment spoke for itself, and yet left-brained thinkers denied the results for many decades.

Fortunately, the world has always had its fair share of right-brained thinkers. The wave characteristics of light did not seem farfetched at all to the early pioneers of quantum mechanics in the 1920s. It was one of these progressive theorists, French nobleman Prince Louis de Broglie, who stated that, in keeping with $E=mc^2$,[6] matter might also be considered to have the properties of a wave. If matter and energy were interchangeable, then matter should logically have the same properties as light energy. Einstein was in agreement since de Broglie's claim complimented the Special Theory of Relativity.[7]

This concept came to be known as wave-particle duality. Light was both a wave and a particle, and since matter and energy were interchangeable, matter, too, had the property of a wave (energy), as well as position (the location of an electron in space-time). Matter and energy were just different aspects of the same thing. The double-slit experiment, it turns out, works for particles (electrons) as well as light waves.[8]

But along came Werner Heisenberg in 1927, declaring that electrons cannot have both energy and location *at the same time.* As soon as an observation is made, the electron has no energy or movement, only location. This was a scientific fly in the ointment, if ever there was one, a fly that has come to be the cornerstone of quantum physics: the Uncertainty Principle. (If this sounds a bit mystical, then it means that your right brain is becoming more than a little curious now.)[9] An electron has the *potential for* energy, and energy has the *potential* to become an electron. Hence we are back to the malleable nature of matter and energy . . . and the supreme importance of potential.

The ramifications of Heisenberg's Uncertainty Principle are considerable, for if we make the mindset of an observer preeminent, then we have turned the universe on its ears. In keeping with the philosophy of the *Tao*, we ourselves have the potential to create, to determine what is real.[10] In the past twenty years, modern medicine has already used quantum mechanics to emphasize aspects of mind-body healing. Dr. Bernie Siegel, in *Love, Medicine and Miracles*, has demonstrated clearly how our moods and emotions, when buoyant and positive, can prolong life and even effect cures of serious diseases, such as cancer.[11] Deepak Chopra, among many others, has reached the same conclusion in Quantum Healing, in which he advocates the idea that thoughts generated in the ocean of consciousness literally manufacture various brain chemicals that, in turn, affect our immune systems either positively or negatively.[12] (The Hindu concept of consciousness is akin to the pregnant void of the *Tao*). Furthermore, Chopra speaks of the placebo effect. If a person *thinks* he or she will get better, even if the medicine administered in a controlled experiment is nothing more than a sugar pill, the individual will more than likely get well. Why? Because the patient is making a quantum observation and determining the reality of biological processes within his or her body.

In considering the impact of the Uncertainty Principle, we should also note a pioneer in thinking strategies, Dr. Norman Vincent Peale, who said in his landmark book *The Power of Positive Thinking* that our attitude determines everything.[13] Negativity and doubt sabotage our development at every stage, while positive attitudes quite literally work miracles, bringing about abundance, health, and positive outcomes in almost any situation.

Uniting Taoist belief with the principles of quantum theory, we can say that there is no hard and fast reality, but rather the *probability* that a given reality will be manifested. *Wu* can become *yu*. Real-time events are elicited from the pregnant void as *chi* is utilized. In quantum mechanics, thoughts or observations are translated into space-time reality. In both systems, it is potential energy that is being realized.[14]

In mind-sciences, we see potential, whether quantum or mystical in nature, as the intentional application of information. From the void come unformatted symbols, which are recognizable. They are then formatted by awareness (consciousness) into meaningful symbols, such as letters, words, or gestures that are ultimately communicated. Whether we choose to apply this information, to format it for the attainment of potential, is up to us. This leads us to the NATI equivalent of the Uncertainty Principle. When we use any of our thirteen intelligences, we create a range of possible meanings. The intelligences form a holistic relationship (natural, innate energy) with each of the other intelligences, and together with their thirteen Polarities, form a range of probabilities, with infinite potential centered at the very heart of the matrix, just as *chi* is centered in the living void.

The Great Restrictors

In the last chapter, I mentioned the Great Restrictors of Fear, Ego, Ignorance, and Self-deception, noting that we must look past the selfish impulses of our individual personalities. In doing so, we allow ourselves access to the pregnant void mentioned earlier in this chapter, a merging of our individual consciousness with the very universe. Any discussion of decoding potential necessitates an examination of the Great Restrictors, which are the main stumbling blocks in achieving healthy goals. Let's look at the Restrictors, therefore, to see how pervasive they are in our thinking. In a later chapter on Polarity, we shall see how weaknesses that result from them can actually be used to enhance the development of potential, of converting *wu* into *yu*. Indeed, recognizing weakness is crucial to the entire NATI system.

Fight or Flight

Approximately 225,000 years ago, during the Paleolithic era, man was a hunter-gatherer, his life filled with the stress inherent in primitive day-to-day survival. Man's primary response to this uncertain existence in a harsh environment was emotional in nature. His sympathetic nervous system evolved what is known as the "fight or flight" response, a biological mechanism still very much present in each and every human today. When faced with danger, a person's bloodstream is pumped full of cortisol, putting his brain on high alert so that he'll know whether to run from danger or confront it directly. This heightened state of alertness raises blood pressure, increases heart rate, and dilates the pupils. It is even responsible for the tingling sensation we get on our skin, particularly our forearms or the backs of our necks, when we are frightened. In general, fight or flight causes a marked increase in anxiety in the human organism. While this response enabled early man to escape danger, its legacy does not necessarily serve our emotional states in a healthy way in the twenty-first century. This same activation of our nervous systems is triggered by hundreds of daily pressures, such as rushing for the bus, being chewed out by the boss, encountering the schoolyard bully, or worrying about paying the rent. There are literally millions of stressors present in modern life. Our lives, in fact, are so fraught with tension that the sympathetic nervous system is frequently stuck in overdrive for much of the day. Evolution has therefore predisposed us to the Great Restrictors, namely Fear, Ego, Ignorance, and Self-deception. In essence, we are genetically hardwired for these, which is another way of saying that evolution has worked both for and against us over the millennia.

Unfortunately, the Great Restrictors are the antithesis of development so that man is frequently a product of rote, mechanical thinking. In so many areas of culture—government, education, and religion, to name just a few—we use an emotional approach to problem solving, one that is based more on the chemicals produced by the reptilian parts of our brains than on logic or reasoning. Paradoxically, we don't solve many problems at all in the long run, although we do succeed all too well in degrading our social systems while eroding our personal health in the process. Mind-body medicine has shown clearly that stress caused by the Great Restrictors is responsible for much of our ill-health, especially heart disease, cancer, and compromised immune system functioning. A vicious circle is then created (a closed system, if you will, with near-lethal feedback) whereby compromised health impairs our ability to function at higher levels or use Directional Judgment to achieve virtue.

The consequences can be seen everywhere around us. For instance, while virtually every American believes that we live in a democratic environment, a limited number of citizens are willing to work in the support of a free and open society. In the 2000 U.S. Presidential elections, only forty-eight percent of eligible voters participated, and turnout for local elections is usually even worse. We seem to have lost sight of Jefferson's timeless admonition that "The price of liberty is eternal vigilance." To believe that our society will continue to preserve freedom given such apathy in the general population is the height of arrogance and self-deception.

Consider the following facts. We are rapidly approaching the point at which fifty percent of the work force will be employed by one of the hundreds of governmental agencies that have sprung up within our nation's bureaucracy. Theoretically, this group could decide to vote *en masse* to effect a change in political or economic policy, with the other half of the electorate having no choice but to go along with the block vote of the cohesive fifty percent. In many ways, this is already happening. The salaries of some local, part-time politicians have been set by a handful of voters, even though the salaries are totally out of proportion to the work performed by those in office. By the same token, activists and lobbyists now seem to have far more influence in governmental policy than scholars, intellectuals, or reformers. A minority of people is influencing the decision-making processes, which belong in the hands of the majority. Perhaps the most frightening aspect of this concentration of power in the hands of a few is that people have grown indifferent towards political processes. They have become discouraged, believing that they can no longer make a difference in the face of big government. Behind this fatalism is a restrictor, namely ignorance, that America is a government by, of, and for the people. This attitude of skepticism is based on emotion, not an understanding of how the Constitution provides for a self-governing population.

Another area in which the unhealthy emotions of indifference and apathy operate is education. Elementary subjects such as world history have either been eliminated or altered by concessions to political correctness. In many instances, discipline is almost non-existent in urban, middle-grade classrooms, and teachers are cowered into submission by student threats or school populations that must pass through metal detectors each morning. In college, the only requirement for a degree is paperwork, the price of admission, or athletic prowess. The result is that both values and academic performance continue to decline in our nation's schools.

As far as religion is concerned, the United States has always been described as a God-fearing nation, so it is no surprise that ninety percent of Americans profess a belief in God. Sixty percent of these belong to a formalized religion, and yet only half of this number practices their faith in any significant manner. Of the remaining thirty percent, there is an alarming number who do not even understand the basic tenets of their faith.

The picture is no brighter when the entire continuum of culture is examined. Corruption in business is rampant, as evidenced by the recent Enron scandal and others. Radical subcultures are evolving at an alarming rate, ones involving cults, mind control, suicide pacts, drug use, gangs, sexual abuse, and hedonism. Reality shows have reduced sex and love to strategies enabling contestants to win millions of dollars. The shows' premises are based on deception, illusion, and rejection, and yet they pass for entertainment, with most viewers never questioning the values that are the foundation for the recent "reality craze."

As Burt Laurel of Laurel and Hardy says, "This is another fine mess you've gotten us into, Ollie."

If one carefully looks at the behavioral patterns that lie behind this "fine mess," one begins to see very non-productive, unhealthy aspects of the human personality. The following represent but a few.

- **Unconcern:** I will stay aloof from situations lest I learn something unpleasant about myself (and/or others).
- **Non-involvement:** If I don't deal with an issue, I can escape dealing with its consequences.
- **Specific Structure Dependency:** If things don't work in a specific manner, I get lost and can't find my way.
- **Superficiality:** Everything is fine the way it is, so I won't jeopardize my comfort (or status) level by exploring new ways of thinking.
- **Elitism:** I am better than anyone else . . . until I need their cooperation and am willing to do anything to win them over.
- **Ignorance:** I will miss the point on purpose so I won't have to deal with an issue.
- **Deception:** What I say is actually the opposite of what I really mean because I am afraid people might see the real me.

- **Introversion:** I hold things in even though I sometimes want to shout or bother someone, but I don't deal with these feelings because I want to avoid confrontation.
- **Violence:** I attack others because I myself may be attacked, and I don't have the ability or power to deal with people, other than physically.
- **Arrogance:** I pretend to know more than anyone, though this is a facade to prevent others from thinking I am stupid.
- **Egoism:** I am the center of the universe.

Every one of these behavioral orientations can be found in the societal "mess" described earlier. The patterns can be found in politicians, students, educators, and ministers—they can be found, in fact, in all walks of life. The tragedy is that people expend a great deal of energy in using these orientations as they try to insulate themselves from pain and hide their true feelings. This is a result of comparative judgment.

There is an entirely different range of thoughts and mental processes, however, which can lead to positive outcomes. Mankind need not be forever trapped by evolutionary patterns that generate the Great Restrictors. The next step in mankind's evolution is neither fight nor flight. It is a willingness to face issues in order to grow and develop. By reaching for greater potential, man can create a new paradigm capable of regulating all of his customs and institutions.

The Great Restrictors—Fear, Ego, Ignorance, and Self-deception—have resulted in the erosion of many institutions, such as government and education, but Natural Thinking & Intelligence offers us a way to improve ourselves and society as a whole. Everyone possesses thirteen intelligences, corresponding to the thirteen phases of cell development. The thirteen intelligences represent a code written into nature, which allows us to realize our potential. They do so by an absolute model, which makes planning, organizing, and taking action a recognizable system. Each of us has these intelligences and we all utilize them everyday. This book is about the science and philosophy of their characteristics and organization.

Summary

Just as it was crucial to find a correct definition of intelligence, it is necessary to understand the nature of potential. One way of viewing potential is through the Taoist tradition, which sees energy existing in a living void that is pregnant with all possibilities capable of manifestation in the world of solid matter. In Western terms, quantum physics says much the same thing in Werner Heisenberg's Uncertainty Principle, which states that an electron can have either mass or energy, but never both at the same time. When we observe an electron, its energy is immediately manifested as a particle. In both Eastern and Western traditions, therefore, matter originally exists as pure potential energy, or possibility. It is this primal energy we tap into and transform whenever we focus upon achieving a goal.

Furthermore, it is crucial to consider the Great Restrictors if we are to convert rigid, programmed thinking into positive outcomes.

PART TWO

THE INTELLIGENCES

CHAPTER FIVE

A + B = C
THE CREATIVE INTELLIGENCE GROUP

$a^2 + b^2 = c^2$
—Pythagorean Theorem

We now move into the definitions of our thirteen intelligences. James Flick, co-founder of DNA, states in his book *Chaos*, that nature is comprised of four basic characteristics, namely Creativity, Organization, Function, and Polarity.[1] These are principles of Chaos Theory. They are also the four basics of Derald Langham's cell model! In a real sense, everything that exists falls within the scope of these four factors. Potential has the ability to create, organize, and function, and it does so precisely because it includes opposite forces within its dynamic. (We will examine Polarity and its place within the NATI matrix in a later section.)

The first three intelligences belong to the Creative Group.

THE FOCUS/AWARENESS CREATIVE INTELLIGENCE

Focus or Awareness is one of the most significant intelligences in that it is, for all intents and purposes, where everything begins. It is connected with all other types of intelligence since everything we do requires some level of focus or awareness. Athletes use an incredible amount of focus to enter what they call a "performance zone," a mindset in which they are able to block out all distractions and engage in peak performance.

Biologically, it is a proven fact that various muscle groups perform more efficiently when the brain has eliminated sensory input not directly related to a given physical activity. In the bodies of athletes with heightened focus, various metabolites are more readily available, and muscle tissue manufactures greater amounts of ATP, or adenosine triphosphate, which is the primary energy source for all muscle contractions.

One indication of the importance and power of Focus is the body's response to imminent danger. Numerous studies have demonstrated that events are sometimes perceived in slow motion during the final seconds before car wrecks, airplane crashes, or any number of accidents that pose a real threat to survival. Although the precise biological mechanisms responsible for this phenomenon are not completely understood, some researchers theorize that the brain becomes hyper-focused, literally kicking itself into a different gear so as to access the greatest amount of information in the shortest time available. Focus is an evolutionary mechanism that aids us in a variety of ways, the least of which is survival.

Awareness

Focusing is concentration, while awareness is consciousness. Although two different factors, they both fall under the same principles which Pythagoras described as will! For the sake of clarity and simplicity, the NATI system adapted Focus/Awareness.

In the early 1960's Professor Eugene Glendin, of the University of Chicago, determined why some therapy was successful and others were not. In a word, the end result was that the successful clients were actively "engaging" in the therapeutic process. They were utilizing more than just logic, or as he put it, "they didn't just stay in their heads."[3]

According to Ann Weiser Cornell, Ph.D, in her book *The Power of Focusing*, focusing is a natural skill that was discovered, not invented.[4] "It is a very broad purpose skill."[5] In biological psychology, awareness describes a human or animal's perception and cognitive reaction to a condition or event. Awareness does not necessarily imply understanding, just an ability to be conscious, feel, or perceive.

Popular ideas about consciousness suggest the phenomenon describes a condition of being aware of one's awareness. Efforts to describe consciousness in neurological terms have focused on describing networks in the brain that develop awareness. Within an attenuated system of awareness, a mind might be aware of much more than is being contemplated in a focused-extended consciousness. It also speaks to the veracity of trusting nature or principles of nature. Without seeing the end result, this translates into "letting go."[6]

Another example of Focus lies in the areas of Open Focus. This concept was pioneered by Dr. Les Fehmi of the Princeton Biofeedback Center in Princeton, New Jersey. Dr. Fehmi is regarded as the dean of biofeedback. His program of Open Focus enables one to participate in Zen like states of mind by "not focusing." He has worked with thousands of clients all over the world, including such personalities as the late Tom Landry, former coach of the Dallas Cowboys. Open Focus actually manifests a heightened state of awareness. This latter factor, Awareness, is a member of the continuum of our Focus Intelligence.

The same basic process, of course, applies to virtually every endeavor we undertake. Whether we decide to write the great American novel, paint a picture, invest in the stock market, or make out a grocery list, every activity we engage in calls for some degree of awareness.

A Firm Belief in Visualizing and Imaging: The *Beliefs/Concepts/Perceptions* Creative Intelligence

The next kind of intelligence in the Creative Group is **Beliefs**, **Concepts**, or **Perceptions**. The intelligence of Concepts refers to how we interpret reality. Everything we are confronted with in life requires that we adopt some concept about it. When we listen to the nightly news, we must decide whether or not we trust the source of our information. We may believe, for example, that one news organization or anchorman is more trustworthy or accurate than another.

Most of us also interact daily with one of the belief systems—government, religion, or education—in our society. It is important to distinguish our beliefs from ordinary perceptions, however, since perceptions may be temporary or lack the intensity of an actual belief or concept. One of the most common indications that beliefs, unlike transitory perceptions, are deeply rooted within us is the old admonition to avoid discussing religion or politics lest we start an argument (and ruin our neighbor's New Year's Eve party). For better or worse, humans can cling with tenacity (and sometimes downright bad manners) to their beliefs or interpretations about revered institutions.

The intelligence of Beliefs or Concepts, however, is not limited to "isms" or doctrines. Again, this creative intelligence refers to how we interpret reality. When Henry Ford began his car company, only a few automobiles were assembled each day. Men worked in small groups of two or three, using parts ordered from other companies. Seeing the popularity of his Model T, Ford moved his operation to Highland Park, Michigan in 1913 and introduced the concept of assembly line production for cars. Delivery of parts to workers via conveyor belt was carefully timed to keep the assembly line moving smoothly. Next, Ford introduced the idea of financing since he suspected that not every family could afford a $600 car. In short, Henry Ford entertained a very concrete belief about the manufacture of automobiles. His interpretation of reality—in this case, societal trends pertaining to travel and finance—convinced him that transportation for everyone was viable in the United States. His belief in the mass production of cars and the need for greater individual mobility led to the concepts of modern assembly-line production and consumer financing. (We'll excuse Henry for also introducing the concept of debt to millions of Americans.)

Finally, Henry Ford most assuredly believed in these concepts with a fervor that transcended mere transitory perception. A byproduct of his invention is an infrastructure traveled by tens of millions of people every day, an infrastructure where fender-benders are regrettably more than just a question of perception. Because of his strong beliefs, therefore, Ford was able to visualize an entire, integrated system of manufacturing, purchasing, and transportation. Further, his belief in capitalism and profit was equally strong.

Belief is also connected to confidence. Further, it relates to imaging, a vital component of potential (something that does not exist, but can be imagined.) In short, belief is an important component of human intelligence, one that drives creativity at a deeply fundamental level.

THE WORLD IS A STAGE
THE *COMMUNICATION/EXPRESSION* CREATIVE INTELLIGENCE

The final Creative Intelligence is **Communication** or **Expression**. In simplest terms, this intelligence refers to how we communicate. There can be no doubt that this is an elemental, innate form of intelligence, for there is little in life that does not bear the mark of how information is presented. Some of the greatest personalities throughout history have been superb communicators. The best example in modern times is probably President Ronald Reagan, known as "the great communicator" because of the effortless delivery of his speeches. While he is not necessarily remembered for the raw strength of his intellect, he remains a popular figure primarily because of the great ease, affability, and enthusiasm with which he communicated his beliefs. It would not be an understatement to say that inability to use engaging modes of expression is sometimes an insurmountable handicap in the political arena. Actual issues aside, one of the greatest criticisms of current President George W. Bush is his lack of smooth articulation.

Winston Churchill and Billy Graham are two other twentieth century figures known for their oratory. It was Churchill, speaking in Missouri on a visit to the United States in 1946, who coined the term "Iron Curtain" as he warned of the expansionist tendencies of the U.S.S.R. Today, almost sixty years later, the term is still recognized, even by generations not yet born when Churchill made the speech.

Clearly, Communication is one of the most basic intelligences, for how often are people impressed not by *what* is said, but *how* something is expressed? Each and every day, we both send and receive hundreds—perhaps even thousands—of communications to the people we interact with, from friends and family to business associates. A simple transaction with a store clerk can run smoothly and efficiently or become an exercise in anger and frustration depending on the communication skills of both buyer and seller. How often have arguments ensued and tempers flared because one of the parties was perceived to speak with aggravation, condescension, or sarcasm? By the same token, people harboring widely divergent opinions can communicate quite effectively when they are skillful at expressing themselves and are able to do so without unnecessary emotion or judgment. Again, what is important is not always the actual content of communication as much as the manner with which it is conveyed. This holds true for other forms of communication as well, such as body language, and other forms of overt and covert behavior. Various postures and poses can convey great sympathy or openness, while others—a rolling of the eyes, a frown, or a simple turning away from a speaker—can convey as much hostility as the spoken word.

Shakespeare said that "All the world's a stage" and he was right. All of us constantly express ourselves in one way or another.

Pythagoras, The Bhagavad-Gita, and Me: The Human Character Formula

Let's now look at a philosophical/scientific basis for naming the three creative principles as we have. The notion actually stems from a human character formula based in antiquity.

Ultimately, Communication goes beyond the style of its delivery, for the character of expression has its basis in the science and mathematics of geometry. In NATI, what you focus on and what you believe about that focus will always equate with how you express it. To put it into a formula, called the **Human Character Formula**, we end up with the following expression of the Creative Group:

$$A + B = C$$

A (awareness) + B (beliefs) = C (character of communication)

The character of our communications can be traced to an ancient mathematical formula known as the Pythagorean Theorem. Pythagoras was not only a mathematician, but also a thinker and philosopher. While his theorem expressing the relationship between the sides of a triangle is still widely known and taught, Pythagoras himself correlated the theorem to his own philosophical thesis of:

will + belief = expression

Pythagoras and his contemporaries were just as interested in the meanings behind a formula as they were in the actual theorem or postulate. This is not surprising when we consider that Pythagoras was part of a much larger philosophical tradition that permeated much of Greek culture.

Looking beyond pure science and its applications was also part of my own search for truth. In my journey to find a unified theory of understanding and intelligence, the logic and reasoning behind certain principles became far more important than the science itself. This led to a habit of visualizing and understanding issues on a visceral and metaphysical level. It was easier for me to understand things in the abstract, which then enabled me to better understand underlying meanings and applications. Indeed, this ability to approach matters abstractly is what gave me the license to alter the Pythagorean Theorem, without corrupting its essence or purity, to A + B = C. (The fact that this formula works impeccably clearly demonstrates its validity.) It works for human systems because science defines nature's principles, which in terms of General Systems Theory carries forward to human systems. Moreover, Pythagoras and his followers developed their sciences through philosophy.

The idea that one's Communication is a product of his Awareness and Beliefs has been validated by psychologists, psychotherapists, and mind scientists time and time again. Problem solving and conflict resolution become much easier when issues are reduced to one of these creative components (or some combination of the three). Marriage counseling, for instance, attempts to make one partner more aware of the other's beliefs in order to lessen non-productive verbal exchanges—otherwise known as fights-in the bedrooms of America. Another important goal in such therapy, by the way, is to make couples aware that *perceptions* rather than facts may be responsible for interactions.

Before moving on to the next intelligence group, it is worth noting that integrated psychology and Eastern philosophy both describe reality using three simple components.

In psychology, the three contributing factors are nature, nurture, and creativity. Nature refers to unconscious instincts, while nurture refers to cultural conditioning. The third component, creativity, is considered to be a drive from the Collective Unconscious (the latter being a Jungian term to describe the consciousness shared by the species as opposed to the limited perceptual reality experienced by any given individual). Specifically, it is a drive toward potential.

In the same way, the Bhagavad-Gita describes three primal components of life: the libido, corresponding to nature; past conditioning, corresponding to nurture; and creativity, which in Eastern thought is the drive toward new concepts attained through the medium of whole consciousness, also referred to as non-local intelligence. This non-local intelligence is the same as Jung's Collective Unconscious.

The relationship of these models to the NATI concept of A + B = C, as well as its ideological Pythagorean counterpart, is perfectly analogous given the fact that they both work. Moreover, it has worked in over 2000 NATI applications over the years. The commonality in all of these systems is that they describe a drive toward creativity and expression, a drive that involves a visceral, intuitive path made possible by the existence of the Collective Unconscious. Just as I was able to better communicate certain underlying meanings after approaching matters on an abstract level, we can all experience enhanced creative expression by functioning at the level of non-local intelligence, or the abstract (for example, being intuitive, open minded, etc.) This is closely allied to intuitive functioning, which will be discussed in detail shortly. It will also lead us to a totally new analytical/decision-quantum-action model of daily reality called the "Potential Intelligences Matrix."

SUMMARY

Mirroring the first phase of Dr. Langham's geometrical model, the first group of NATI intelligences is the Creative Group.

Focus (or Awareness), the first intelligence, connects with all other intelligences since everything we do requires some level of awareness.

The next intelligence is Beliefs (or Concepts). Everything we encounter requires that we adopt some kind of belief about it. Beliefs are deeply rooted within us as opposed to perceptions, which are usually transitory in nature.

The final intelligence in this group is Communication (or Expression). This is an elemental intelligence since most of the information we gather in life is expressed in some manner, either verbally or through action.

Together, these three intelligences form the Human Character Formula, which is A + B = C. In other words, what we focus on, plus what we believe about that focus, will always equal the character of our communication (or what we are capable of becoming, since the acts of "becoming" and "expression" are closely related).

Problem solving and conflict resolution are much easier when issues are relegated to these creative components.

Chapter Six

Welcome to The Organization: The Organizational Intelligence Group

Lay down all thought,
Surrender to the void . . .
—John Lennon

Webster's Dictionary defines the word "organization" as "any unified, consolidated group of elements; a systematized whole."[1] This definition, brief as it is, adequately introduces the next six intelligences, for they comprise the **Organizational Group**. They are indeed a consolidated group of elements that culminate in the idea of wholeness or synthesis (i.e., they will ultimately subsume all thirteen of the intelligences in the NATI matrix).

If the Creative Group reflects the cell's polar pulsing in Derald Langham's geometric model, then the Organizational Group reflects the wave motion of cellular development. Just as the triadic nature of the creative intelligences is demonstrated by the three axes of polar pulsing in the initial development of the cell, the next grouping corresponds to the second stage of development, which displays six separate though interrelated points (i.e., the six faces of a cube). Once again, in General Systems Theory terms, cellular development, as a scientific principle, is analogous to human systems, in this case NATI organizational principles\intelligences.

It is also worth noting that the structure of DNA structure has six facets. DNA is made up of four nucleotides forming any number of base pairs. But the form it follows is comprised of a six-part model! According to the findings of Bucky Fuller and his contemporary Derald Langham, crystal structures in nature, containing a quality of "six-ness," all relate to organizing.

Playing Tennis With A Net: The Organizational Intelligence Of Laws / Models

The first organizational intelligence is **Laws** or **Models** and relates to several notions of rules, laws, and models. This intelligence refers to the adoption of an image or picture of what works. While society can be indifferent to modeling, NATI prefers to say that there is a certain amount of truth in the cliché that "imitation is the sincerest form of flattery." Indeed, modeling is severely underrated as a way of improving human functioning. There are literally hundreds of how-to books about writing that have been published, and virtually all of these manuals spurn imitation as a way of developing a unique narrative voice and style. This advice to shun imitation betrays an ignorance about the creative process, for it is precisely through imitation of *several* styles that a writer can synthesize a new voice, taking a little from each author studied or imitated until various stylistic elements add up to a new kind of prose that is totally distinctive. This kind of modeling actually goes back to the Greeks, who believed that the imitation of models was the best way to attain perfection. Even in more recent times, however, education emphasized imitation as a way to perfect writing skills. In composition classes, following proven methods of organization was considered a sound way of honing a person's writing talent. Sadly, modeling has fallen into disfavor under new curriculum guidelines that mistakenly think individuality is unattainable through the study of tried-and-true methods. Students are encouraged to simply experiment in a vacuum.

The same also holds true for creative writing. Alluding to the writing of poetry lacking rhyme or meter (called free verse), Robert Frost said that before one can play tennis without a net, he must learn how to play tennis *with* a net. Before attempting to write new or experimental verse forms, it is necessary to study (and practice) classical models of poetry.

Business is another area where the use of models is critical for efficient functioning. Businesses utilize models for numerous applications. They implement laws as constructs or organizing principles. In terms of management, most businesses are built on the concepts of authority and chain of command. (Later in the book, we will discuss progressive models of business that allow worker input in a less hierarchical structure.) Businesses also rely on economic laws of free enterprise, profit, and supply and demand. Many corporations have research and development sections that produce new products based on existing models, trends, or technology.

The plain fact of the matter is that Laws and Models teach us what works. While it may sound simplistic to think of Laws or Models as a form of intelligence, consider something as rudimentary as making a decision to eat a balanced diet. People who are overweight cannot shed pounds without a safe diet plan, supplemented with exercise. While fad diets and weight loss pills flood the market daily, the most proven method is to cut back on saturated fats while exercising on a regular basis in order to burn excess calories. Effective models of dieting show us what works and what doesn't in our efforts to maintain health. We can choose a fad diet or amphetamine-based pills, but the benefits will either be short-lived, dangerous, or both.

Another example is bodybuilding. If we attempt to strengthen or tone our muscles with incorrect exercises or by using the wrong equipment, we stand a considerable risk of injuring ourselves. But working intelligently and systematically—using exercise models prescribed by a weight trainer—allows us to reach the goals of increased strength or improved health. It is a matter of simple common sense to consult established norms before engaging in almost any activity. The fact that we often lack such common sense in matters of exercise is why chiropractors make such a good living!

It is by focusing on pre-existing models that we can be assured of events or outcomes. As is the case with writing, the paradox is that we can often create new models only after we have envisioned older ones. The latest model of any car is an improvement on existing models that date back to Henry Ford's original design for the first automobile. But where, we might ask, did Ford obtain the *original* design? The answer is perfectly straightforward: he created the automobile by modeling his ideas on existing principles of motion and combustion, and it is no accident that Ford called his first highly successful car the *Model* T. Ford seems to have been acutely aware that his design was a prototype that would go through endless redesign over the years.

Even the process of biological evolution is based on the concept of Models. The process of natural selection increases the chances of survival for some animals within a species that has already experienced some degree of success in adapting to its environment. The very mechanism of evolution is based on the concept of using what has already been proven to work.

We can even see modeling taking place in the development of a human fetus. A biological and philosophical axiom is that "ontology recapitulates phylogeny." This principle states that the development of an individual organism reflects the development and evolution of the entire species to which it belongs. With the human fetus, for example, we see that the growing embryo exhibits many characteristics of earlier, more primitive forms of life from which it evolved. In early weeks, it is essentially amphibious in nature.

In short, nature's own economy and wisdom, aimed at potential growth, seems to place a very high premium on models.

The Devil Is In The Details: The *Details* Organizational Intelligence

The next organizational intelligence is **Details**. This is a straightforward but essential intelligence because it has pronounced implications for the synthesis of the thirteen intelligences into a holistic matrix wherein all of human life functions. The idea of Details relates to the individual's ability to perceive parts, specifics, levels, or separate items in the context of a larger system. It is related to the function of General Systems Theory—the breakdown of a system, object, or idea, into its constituent parts. (We will examine General Systems Theory in considerable detail in later chapters.)

The importance of Details is evident in everything around us, and yet we probably don't recognize this intelligence for the same reason that that we do not see the forest for the trees. Virtually everything in existence is comprised of components, although most of us generally focus on the whole rather than constituent parts. The same is true for issues or ideas. We focus primarily on a general concept rather than the details (or nuances) that may underlie a particular idea.

It's only natural that we do this, of course, for we would never get through the day if we stopped to examine every part or detail we encountered. All matter, for example, is made of atoms, which themselves are broken down into smaller particles, such as protons, neutrons, and electrons (which are broken down into even smaller particles in the dizzying subatomic world described by particle physics). As creatures in the macro-world, however, we do not normally focus on the chemical nature of reality. We are more than content to deal with three-dimensional objects such as tables and lamps and chairs.

The communication of information is another excellent example of Details. Computer information results from a binary language of 1s and 0s that form information stored on hard disks in the form of quantifiable parts known as bytes or megabytes. There literally could not be any form of communication without the intelligence of Details. The more detailed our communication, the more comprehensive and intelligent it is.

Technology, especially since the Industrial Revolution, could not exist without the intelligence of Details. Machinery is made of components that may be as simple as cogs, wheels, nuts, and bolts, or as complex as circuits, wiring, and computer chips. In one sense, the history of evolution and technology is the story of how man became a toolmaker and used more and more complex parts to build larger and more complicated machines.

As mentioned above, most people do not focus on Details . . . until something breaks down. Otherwise, we neglect components or details unless our profession necessitates that we focus on the composition or make-up of something. Doctors, electricians, computer technicians, carpenters, and architects (to name just a few) are examples of professions that must come to terms with Details.

We shall see in later chapters that certain aspects of holography and systems theory demand that we look at how components are synthesized, but without the existence of specifics to begin with, any discussion of holism would be meaningless. For this reason, Details is one of the thirteen fundamental intelligences of the universe.

Bonsai Equals Productivity Squared: The *Order/Processes* Organizational Intelligence

The next intelligence is **Order** or **Processes**. In its simplest terms, this intelligence encompasses all sequential patterns of action and manifestations (i.e., all procedures). It is literally how we do things; it is the procedures we follow, the patterns we utilize.

Daily routines are examples of processes we engage in to accomplish various tasks. Every morning, many of us get up and go through certain rituals—grooming, dressing, and eating breakfast—so that we'll look presentable in the workplace. We then drive to work or use mass transportation that runs according to a certain schedule so that we will arrive at work on time. Throughout the day, we then adhere to more schedules in order to accomplish a quantifiable amount of labor. This is done for the dual purpose of furnishing society with goods and services, as well as providing a personal income, making possible, among other things, the continued practice of the initial processes of getting up, grooming, and going to work. The processes in this example are also a function of order in that they stem from a certain structure imposed by society and, as they are carried out, help to maintain that order. When procedures work efficiently, they create a positive feedback loop that extends from, and leads back to, order.

It is not surprising that many people, if not most, complain from time to time about the predictability of their lives. The parent laments the seemingly never-ending routine of cleaning house, driving children to school, preparing meals, and then jumping in a minivan to pick up the kids and bring them to afternoon soccer practice. Similarly, it is not uncommon to hear a wage-earner express frustration at being caught in the "rat race," a cycle that all too often seems to exist only for the purpose of perpetuating itself. Certain repetitive patterns can literally cause depression in some people.

By contrast, NATI approaches procedures with a mindset that works in harmony with nature, enabling people to transform frustrations over routine into action and achievement. The quintessential example here is the way in which the Japanese have integrated holistic health into their workplaces and factories. Japanese corporations schedule frequent breaks during the workday for rest and exercise. Many companies even permit time for recreational activities or hobbies such as bonsai, which is a novel approach inasmuch as bonsai itself requires focus, concentration, and order, and yet it is perceived to be relaxing.

This more personal approach to employee-management relations may seem strange to the Western corporate mindset, but the reality is that the Japanese are using various NATI intelligences to enhance productivity by enabling the worker to realize greater potential. In this sense, there is a positive behavioral feedback loop within the larger manufacturing feedback loop of order-process-order. The Japanese model indicates that predictability need not constitute an imposition on our personal growth.

In the NATI universe, therefore, it is important that we understand that there exists a natural order, one of the key components of which is predictability. When we are able to grasp the inherent order in a system, we are often able to predict what is coming next. In climates where the four seasons are discernible and show the classic variations in temperature and precipitation, it is easy to know what weather pattern is coming next. Summer follows spring, fall follows summer, winter follows fall, and spring follows winter. Knowing this pattern enables people to plan their lives in many ways. In business, department stores know when to put season-appropriate lines of clothing on display. In healthcare, professionals know when to prepare for flu season or the treatment of sunburn. These are all very positive aspects of order, process, and predictability.

David Bohm, one of the greatest theoretical physicists of his generation, believed that order was also implied by the ability to perceive the difference of similarities as well as the similarity of differences. This may seem like a riddle or word game, but it is actually a methodical way of understanding the elemental aspects of patterns and is a key component of our Potential Intelligence Matrix, which is discussed in a later chapter. The Special Theory of Relativity is an example. Einstein showed that matter and energy were equivalent. Matter is similar to energy, and yet different from it. Likewise, energy is similar to matter, and yet different from it. Ultimately, though, the similarities and differences are part of the very same feedback loop of order-process-order. Matter is converted to energy, which is then converted back into matter. Fire releases heat, which is then absorbed by natural elements or man-made objects. Conversely, energy is converted temporarily into matter, only to become energy again, as when sunlight drives the process of photosynthesis, creating plants, which then release energy into the atmosphere during their growth cycle in accordance with the Second Law of Thermodynamics.[2] A system of order is always preserved through these scientifically predictable processes.

Procedures can describe incredibly complex degrees of order. Although patterns unquestionably apply to the mundane—the manipulation of everyday objects or the recognizable sequences we use to accomplish various tasks—they are equally valuable in understanding scientific, psychological, and philosophical ideas.

In short, the intelligence of Order or Processes shows us the steps involved on the path we travel.

The Tao of Assessing: The *Measure* Organizational Intelligence

The fourth organizational intelligence is **Measure**. It is analogous to prioritizing or assessing issues or events. It pertains to the depth of an issue, its significance or priority. We see this intelligence at work all around us, both at work and play. In baseball, the speed of every pitch is measured by a pitching coach. That person relates the information to the manager so that he may assess a pitcher's performance from inning to inning or game to game. A club's staff also evaluates the pitcher's ball-to-strike ratio, the number of walks or hits he gives up per inning, and the movement of his pitches such as curve balls or sliders. Pitching assessments are then added to judgments about hitting, fielding, and the number of team wins to gauge the club's overall performance. Depending upon the team's success or failure, priorities may be changed. The batting order may be rearranged, the pitching rotation may be altered, or players may be traded to different teams.

Assessment is also amply demonstrated in the area of politics. A responsible elected official should always ask what impact a proposed law will have on his constituents. While in the committee stage, most legislation is examined to see what segments of the population will be affected by its passage and whether or not procedures are in place to enforce a law if enacted. Not all laws, of course, are deemed appropriate or just. Any judicial interpretation of laws after they are passed constitutes a very formalized assessment. Indeed, without the intelligence of assessment, there could be no system of checks and balances whereby each branch of government evaluates the performance of the other two branches.

In recent years, political assessment has become extremely sophisticated with respect to polling, focus groups, and demographics. An unfortunate outgrowth of this trend is that politicians are sometimes more worried about how an issue will affect their careers rather than the voting public. This worry is really a manifestation of the Great Restrictors of Fear and Ignorance, which have undeniably eclipsed integrity and honesty in the modern political arena.

Prioritization is another way of expressing this kind of intelligence. When we decide which patterns or procedures to employ, we are engaging in fundamental assessment. A general might feel that it is appropriate to apply the entire spectrum of military assessments to one theater of operations before another. In government, it might be necessary to pass a law requiring taxation before enacting laws on how to distribute collected revenue. In everyday life, we must all select which appointments, errands, or chores take precedence over others.

All assessments, in the long run, must be viewed as relative. In fact, relativity is an important aspect of Measure or Assessment. Interestingly enough, Einstein's Special Theory of Relativity keeps only the speed of light as a constant, while measurements of time and mass vary according to an observer's position relative to this cosmological constant. Relativity is a key factor in the establishment and implementation of the Potential Intelligence Matrix.

In one of the most famous thought experiments in all of physics, an astronaut is visualized as moving away from the earth at a velocity near the speed of light. When he returns to earth, he may perceive that only ten years have passed, although everyone he knew has died since over five hundred years have passed by earth's standards. Length is likewise shortened since a vessel approaching light speed comes closer and closer to becoming two-dimensional. Similarly, the Special Theory of Relativity says that objects become more and more massive as they approach the speed of light, so that an astronaut traveling at high velocities would weigh thousands of pounds, although he himself would not be aware of any changes. He would feel no sudden impulse to diet!

In NATI, we say that relativity is therefore subject to the law of measurement. Because relativity contains zero and infinity and everything in-between, all extremes of measurement must be included in a relativistic universe. In *The Looking Glass Universe*, authors John Briggs and David Peat cite David Bohm as saying "The Laws of relativity are laws of an average not absolute laws".[3] In other words, we are left with an elemental bit of logic that states "If this, then that." One thing is contingent upon another. (There is another bit of irony here: relativity itself can be considered absolute! It then becomes a member of natural law.)

It is this one short word "if" that holds all the cards in the hand of Measurement, for we must remember that quantum theory is subject to Heisenberg's Uncertainty Principle. Remember that an electron cannot have both energy and position. It is the position of the observer that quite literally determines everything. (In other words, it is one's perspective or position in the field of potential that dictates an action.) *If* an electron appears to have energy, it exists in a cloud form, with only the probability of location. In fact, it is in *all* possible locations at once, a small subatomic particle on an amphetamine joyride. *If*, on the other hand, an electron appears to have position, it is like a ship dead in the water, without energy or momentum. It has been sent to rehab to get over its nasty addiction to speed.

To repeat one of the sayings of the *Tao*, "The Tao is mysterious, unfathomable; yet within is all that lives" (Tao 21).[4] Relativity and the quantum theory of probability tell us that there are deeper currents beneath ordinary events. In the all-encompassing system that is NATI, we may apply the intelligence of Measure to intricate systems that would otherwise be indeterminable chaos. Measure is critical if we are to transform the living void, filled with infinite potential, into recognizable patterns of reality. The better one's ability to measure, assess, or prioritize, the more accurate his discernment.

A House Of Mirrors: The *Reflection/'Mirroring'/Feedback* Organizational Intelligence

The fifth intelligence in the NATI matrix is **Reflection**, **Mirroring**, or **Feedback**. This is an extremely visceral intelligence. It is derived from the age-old oriental notion that whatever bothers or disturbs us to any significant degree is really some internal factor being reflected back to us. In other words, whatever unsettles us concerning another is actually something within ourselves that is wrong or incorrect and needs attention.

A good marital therapist will always help patients explore the possibility that complaints about their partners may originate because of some inadequacy inside of themselves, not their husbands or wives. A basic tenet of psychology is that people have a tendency to project their own unhealthy behaviors onto others, in whom they then see the behaviors mirrored, albeit they are completely unaware of this process.

Husbands and wives often claim that they do not receive enough attention (or the right *kind* of attention) from their partners. If a husband or wife feels neglected, he or she may well be neglecting something important in his or her spouse. A husband may not be getting enough credit for the time he spends at work, while a wife may not be receiving enough praise for her role as mother, homemaker, or provider of a second income.

Two excellent examples of the phenomenon of Mirroring involve my good friend and colleague, Terry Anderson. Terry was held hostage in Beirut from 1985 to 1991 by Islamic militants demanding the release of seventeen Shi'as convicted of bombing the French and American embassies in Kuwait. In his book *Den of Lions,* Terry writes of tension that existed between himself and another hostage.[5] After Terry precipitated a confrontation between the two, the other hostage left the cell to wash up. In the absence of the alleged troublemaker, Terry urged his fellow hostages to do something since his nemesis was, in his opinion, being obstinate and inflexible. Much to his surprise, the other hostages told Terry that he himself was the one being obstinate and inflexible and that he was as much responsible for the tension as the other hostage, if not more so. In his book, Terry compares his surroundings while a captive to "a house of mirrors."[6]

Many people immediately dismiss this intelligence because it hits too close to home. Even Terry Anderson rejected the idea of Reflection and Mirroring at first. It wasn't until I reread his book and discussed it with him that he began to give it credibility.

The same thing happened when Terry and I conducted a seminar in Rye, New York in 1998. After explaining the science of Mirroring to the group, I was instantly challenged by several people who attempted to refute the concept. One woman in particular adamantly rejected the entire thesis. I then did what I always do in a situation like this: I looked into the mirror. I asked the woman to give me an example of something that really bothered her, preferably something recent. She related a situation concerning a new job she had just begun, one in which several of her fellow staff members were treating her with arrogance and disrespect—or so she said. Within ten minutes, she openly admitted that the very first time she walked through the door of her new place of employment she had conducted herself in an arrogant and condescending manner.

To reiterate, this is a very visceral science, and I have never seen it fail to provide a valid insight. As to why it exists or what metaphysical meaning it holds, I have my opinions. My own belief is that God, or a God-force, wove it into the fabric of the universe for the purposes of development.

Development, as we have seen throughout this book, is a universal principle and should be our ultimate focus because consciously or unconsciously, we are all striving for it, and Mirroring exists to show us the way. (One might be inclined to say that the God-force seeks total integration of the universe by way of the principle of development.)

By using this intelligence, we can become better human beings, more sensitive and aware of the things we do in our lives.

Granted, it is a difficult concept to deal with, for nobody is perfect—only the path of development is. People who reject this notion of mirroring usually don't accept responsibility very well because it exposes them to their inner selves and this frightens them.

A Unified Theory Of Everything: The *Wholeness/Synthesis* Organizational Intelligence

The sixth and final organizational intelligence is **Wholeness** or **Synthesis**. This is the ability to synthesize information and place it into a larger system. It is the ability to perceive unity and integrate key components into "the big picture." The world is positively replete with examples of wholeness.

In baseball, an ordinary game consists of nine innings. Partial games are accepted into the record books, but only if five innings have been played. The rationale for this rule is that more than half the game must be played so that it may be rounded mathematically to a whole game. The concept of wholeness is implicit even in our national pastime.

In classical music, symphonies have four movements. One of Shubert's symphonies is called *The Unfinished Symphony* because it only has three movements. Archivists could easily have ascribed a catalog number to the composition, but wholeness is so much a part of human thinking that music lovers deemed it important to note the incompleteness of Shubert's work.

All beehives are comprised of three types of bees: a queen, workers, and drones. The duties of the worker are to construct and guard the beehive, repairing it when necessary. The drone exists only to mate with the queen, who lays eggs to supply the hive with workers, drones, and a future queen for another hive. These three functions are so inextricably tied together that the beehive, though made of parts, cannot exist outside of its holistic nature.

Insofar as human nature is concerned, synthesis and wholeness relate to integration, or "the macro." Being accepted into a group, for example, represents wholeness. In NATI, when one has mastered all four Functional Intelligences, they are said to be whole.

We have also seen this intelligence in RIKU, whereby raw information is assembled into meaningful information, such as letters becoming words and then books, or musical notes becoming melodies and then symphonies.

In short, we see this intelligence in both nature and human institutions. The idea of wholeness has even crept into popular aphorisms such as "He can't see the forest for the trees," meaning that a person can't see "the big picture" of the forest because he is looking at the individual trees that make up the forest to begin with. The saying implies a fixation on Parts or Details. This would be tantamount to Henry Ford standing on the floor of his factory, totally flummoxed, while looking at a hundred auto parts, asking "Where's the automobile?"

Some physical structures simply cannot be fully understood in terms of individual parts. Buckminster Fuller's geodesic dome is a prime example. The geodesic dome is a structure made of interlocking hexagons that form an overall dome shape. There is equal weight and force distribution throughout the entire structure of the geodesic. A blow to one portion of the dome results in a crack on the opposite side, exactly one hundred and eighty degrees away from the impact. This is one part affecting another part of a whole.

Organization And Quantum Unification

As we have done in our discussions of the other intelligences, let us now turn our attention to science. We have seen that, according to quantum mechanics, there is a wave-particle duality in nature. Light, and by extension matter (according to de Broglie), may be regarded as both particles or waves, matter or energy. The exception is Bohr's Copenhagen Interpretation of quantum theory, which emphasizes the collapse of the wave function upon observation, so that the wave immediately reverts to the more stable Newtonian picture of reality (particles of matter). What is important, however, is the existence of subatomic particles as *probability.*[7] Implicit in quantum theory is the basic oneness of the universe. This then enables us to connect matter and energy as components of potential. It is not farfetched to then project

potential as the basis of a "Unified Field." The issue of a unified theory of potential is another key theorem of the Potential Intelligence Matrix. True, the act of observation results in the discernment of a particular reality. We cannot debate Bohr or the Copenhagen School on this point, but we must remember that the Copenhagen Interpretation, with its collapse of the wave function, is merely describing the Taoist concept of *yu*. From infinite potential comes the reality we create by observation. Wholeness, however, always points us in the direction of *wu*, the pregnant void wherein all potential and cosmic unity exists to begin with. The manifestation of quite literally *anything* comes from nature's inherent oneness, its all-pervasive unity.[8]

We are essentially talking about a "Unified Theory of Everything," and it is worth noting that Einstein himself grew sour on quantum theory because it could not unite gravity with electromagnetism. While he could not deny the mathematical correctness of quantum equations, Einstein found the notion of uncertainty to be unsettling in the long run. While a professor at Princeton, he declared, "God does not play dice with the universe."[9]

But what if uncertainty is actually a continuum? It would establish itself as probability! It may be probability beyond our comprehension, but it is probability nonetheless. With the thirteen intelligences, this is precisely what occurs! An issue—any issue—is connected with one, several, or all of the thirteen. Clarity then ensues since the answer always lies somewhere within the thirteen principles.

Many readers may feel the same way as Einstein regarding God and uncertainty, but there has been an exciting development in science in the past two decades that is only now receiving recognition: string theory. String theory predicts that all subatomic particles are made of small bits of vibrating energy, visualized as tiny strands, or "string." According to some scientists, such as Brian Greene, author of *The Elegant Universe*, string theory may be the missing piece to the Grand Unification Theory that science has made its Holy Grail ever since the emergence of quantum mechanics in the 1920s.[10] If proven true, this will be an especially exciting idea since we would once again be face to face with pure energy as the basis of all creation. The great void of potential spoken of by the *Tao* already underlies everything that we see, hear, feel, and touch, but with string theory, the pure energy of creation could be even closer than we have heretofore imagined. In NATI terms, strings could actually be pictured as microscopic closed universes strung together at different harmonic frequencies. This notion represents the closed system of NATI acting within an open system. For example, there can be any number of NATI "units," each at work separately, but all are open to each other's influence.

Summary

The Organizational Group of intelligences mirrors the cell's six aspects of wave development.

The first intelligence here is Laws (or Models) and refers to adopting a picture or model of what works—a profile, if you will.

Details (or Parts) relate to an individual's ability to perceive the details or various parts, levels, or separate items of a larger system. It is also related to analysis, or the breakdown of an object or system into its component parts.

Another intelligence, Synthesis (or Wholeness), is the ability to re-integrate parts into a unified whole. It connects with the notion of "the forests and the trees."

Order (or Processes) refers to the various procedures we employ to get things done. Although procedures often imply routine, they make it possible to grasp complex order within a system.

Measure is the ability to assess, judge, or prioritize events or issues, actions we do unconsciously each and every day. Quantum theory tells us that this intelligence is crucial, as demonstrated by the observation of an electron according to Heisenberg's Uncertainty Principle.

The final intelligence in the group is Reflection (or Mirroring) and is very visceral in nature. The principle behind it is that whatever bothers us in others is actually a projection of something that needs attention within ourselves.

The following represents some of the analogous terms on what may be considered a continuum for each intelligence.

ORGANIZATIONAL

• Formative • Structure

MODELS	ORDER	PRIORITY MEASURE	MIRROR/ FEEDBACK	TOTAL/ WHOLE	DETAILS
Rules	System	Levels	Reflection	Whole	Components
Form	Perimeters	Degrees	Complementary	Unity	Individual
Laws	Format	Relatives	Response	Entirety	Segments
Identity	Pattern	Emphasis	Interaction	Overall	Separate
Support	Discipline	Intensity	Resonance	Collective	Unit
Basics	Process	Significance	Feedback	Group	Sector
Agenda	Procedure	Comparison		Composite	Segregate
Models		Judgment		Continuous	Subdivision
Principles		Probability		Generative	Parts
		Infinite		Connected	
		Dimensions		Integrate	
				Synthesize	

WHOLE-TOTAL The highest level of understanding. The more types and styles you can embrace, the better you are. The greatest intelligence is when every part has a fit in the issue. One can see the entirety. Everything is recognized for what it is. It is the viewing of the entire picture with all one's parts.

MIRROR-FEEDBACK We are constantly surrounded by reflections of ourselves, especially the negative. Feedback is a reactionary method of knowing and learning. If it is experiential (vs.insightful/qualitative, etc.) it may well miss long-term knowing. It reflects the problems on the path to an objective.

ORDER-PROCESS These are processes, order, cycles, steps, etc. in any structure, including chaos. This is the place of paradigm shifts, the path one chooses. At some point in a process, underlying concepts clash and hidden patterns emerge. Patterns are mostly rote and should be analytical. Information (intelligence) changes patterns. Processes are steps toward an objective.

MEASURE-PRIORITY This is judgments, priorities. They shift and ebb according to our focus and beliefs (ethics). Relativity is perhaps the greatest, most valid measure. By matching functional and creative principles with core dynamics, a true sense of priorities emerges. Interesting concepts emerge as well.

DETAILS-PARTS	The parts or details are implicated with an issue and are the segments comprising the totality. They are bytes of information necessary to accomplish an objective.
MODELS-LAWS	The laws in question are at issue here. What rules or structures are in place or are to be followed. Imitation is another interpretation here. They are examples of what can or should be achieved or followed. These are the standards for adherence to an objective/action (focus or concept).

Chapter Seven

The Functional Intelligence Group

The final group of intelligences is the **Functional Group**. A portion of the philosophical basis for this branch of intelligences is also to be found in antiquity. The ancient Greek, Hebrew, and Oriental traditions spoke of them at length.

There are only four ways in which the human mind can function: physically, mentally, emotionally, or intuitively. This group corresponds exactly with the Neo-Platonist school of thought, which originated in the sixth century and combined Plato's belief in absolute beauty and perfection with oriental mysticism. It describes man in terms of body, mind, feelings, and spirit, and once more we see that NATI corresponds to Greek models of thinking and intelligence. The Functional Group describes how we think, feel, act, and use intuition to express ourselves and obtain our goals.

The Functional Group may also be derived by examining the evolution of various schools of psychology. Sigmund Freud's theory of personality, for example, was based on his belief that all motivation derives from the human being's unconscious sex drive. It was Freud's belief that a person's relentless pursuit of pleasure was the underlying motivation for virtually all human behavior. There is no denying that the libido is a powerful motivator of human action, for it is directly tied to the idea of species survival. Nevertheless, other psychologists thought Freud's view too limited in scope.

During the same time period, psychoanalyst Alfred Adler advanced a competing theory of personality development.

Adler's theory emphasized a person's mental prowess. He said that man had an innate desire to know things for the purpose of predicting and controlling them. What Adler's theory amounted to was this: humans need to establish power, which is done through cognition, or thinking. This was in stark contrast to Freud's notion of a "libido run rampant" inasmuch as the sex drive is based in the reptilian part of the human brain (located near the brain stem), which seeks to satisfy needs such as survival and hunger. Cognition, on the other hand, is carried out in various areas of the cortex—the proverbial "gray matter."

It was Carl Jung, originally a protégé of Freud, who synthesized the views of Freud and Adler while simultaneously augmenting them. (It should be noted that this kind of synthesis is a classic example of the intelligence of Wholeness.) Jung concluded that everyone was born with four distinct ways of interacting with the world. These four kinds of interaction correspond precisely with the intelligences we have named in the Functional Group.

Each function also corresponds to a personality type. Studies over the years have shown consistently that identifying our personality type is analogous to recognizing how we are motivated to achieve various results. This premise led Katherine and Isabelle Briggs to develop, in conjunction with career counselors and college placement officers, the Myer-Briggs Type Indicator.[1] This psychological inventory is administered to millions of people annually to help assess how a person functions, which in turn can help determine possible career paths. As with the NATI system, it does not seek to determine good or bad, right or wrong. Rather, it is used to help people choose the path to their potential.

Trapshooters and Golfers—Manifesting Potential Under Pressure: The *Physical* Functional Intelligence

The first Functional intelligence may seem to be too obvious to be regarded as an intelligence at all. It is the **Physical**. This intelligence refers to matter, the visible manifestation of potential energy. The significance and nature of physical intelligence was never clearer than when I first started with NATI back in 1981. At that time, I was developing a trapshooting enhancement program for myself and some of my business colleagues. We were all quite heavily into shooting on a competitive basis at the time. It became apparent to me that the Human Character Formula (A + B = C) was highly effective in developing our physical skills, not only in the area of trapshooting, but also in other endeavors such as golf. It showed me that the amount of concentration focused on the task at hand had a lot to do with the degree of success realized, which is the Awareness intelligence. In addition, the ability to believe in yourself, have self-confidence, which is the Beliefs intelligence part of the equation, also came into play to a large degree. It never ceases to amaze me, as it did my colleagues all those years ago, how we tend to defeat ourselves by our own limiting perceptions. You can place yourself on the practice putting green and make one three-footer after another without missing a single one. However, approach that same putt during a competitive round and it's the same as walking across a wooden plank a hundred feet off the ground. It takes on a much greater significance. Success in sports becomes largely a function of how well one focuses on the task at hand, as well as the belief in one's ability to achieve that end.

If it ended there, it would be all too simple, but life is not without its pressures, regardless of how successful or capable one is. I found that developing one's focus certainly had a large impact on performance, as well as confidence and visualizing successful results, but there was a great deal more than that. There were, in fact, the other ten intelligences that played a part in successfully dealing with one's physical abilities, whether it be in sports or any other area of physical activity. Time and time again over the years, experience has taught me that the organization of one's undertaking is just as important as the planning or creative aspects. Knowing what to do and when to do it is equally as important. The processing of data is best served when it is consciously organized. NATI helps us to take innate processes from nature and increase our effectiveness by virtue of conscious understanding and then application.

One of the most amazing things about Physical intelligence and undertaking particular physical scenarios is that the moment one thinks he or she has something mastered, often everything seems to fall apart. The 72 on the golf course one day can easily turn to an 89 the next day without any material changes in the playing conditions. For example, professional trapshooters and golfers can perform brilliantly one day and perform horribly the next day. This is true of all physical undertakings.

I have come to equate this phenomenon with the collapse of the wave function, which, you may recall, is part of the Copenhagen School of quantum mechanics. When an observation is made, the probability wave collapses into an actual event. Analogously, the minute the brain understands the entire paradigm, it seems that we expand into ever-deeper levels of experience. What then develops is a larger paradigm that incorporates the former paradigm. What also tends to occur when one is under pressure is that the new paradigm does not hold up because the recognition and information involved in a new way of thinking has not been fully realized or accepted by the individual. Consequently, the person falls back to a level or paradigm that he or she "is comfortable with." Sometimes, this reverting is a free-fall and brings us back to previous levels. Sometimes these regressions are far below our typical functional capabilities. When trapshooting, for example, the brain organizes billions of neurons in their networks. They are organizing in accordance with the six organizational intelligences. The greater the ability to concentrate and image, the better and stronger the organizational capacity of the neurons and their networks will be. There are, however, limits to this ability to organize by any measure. Accordingly, I have discovered that the best way to retain organizational confidence and focusing power is by adhering to fundamentals and abstractions. An example of this is listening to music or visualizing colors floating throughout one's brain. This brings us to levels of abstract thinking that enable us to bypass the short circuits of restrictive organizational patterns. Applying virtue here can also work, as you will soon see. For instance, some of the best performances have come with complete focus on "my responsibility to be better, to develop."

Throughout the book we discuss the four Great Restrictors: Fear, Ego, Ignorance and Self-Deception. In pressure situations, all four are frequently at work, but, for the most part, the negative emotion of excessive or inappropriate fear is the greatest. Therefore, releasing physical potential can also relate with Emotional intelligence and Polarity. Being emotionally positive, or charged, will go a long way towards releasing potential. Whether one is picking up a putter to sink a putt or using a shotgun to hit a target, the Human Character Formula, A + B = C, plus the Organizational intelligences, are always at work. Understanding what the mechanics are, even in general terms, helps to uncover and expand potential. Some vocational examples focusing on Physical intelligence are athletes, surgeons, mechanics, architects, tradesmen, make-up specialists, pilots, artists, etc.

Raw Emotion, Well-Done Performances: The *Emotional* Functional Intelligence

To say that some people function emotionally might be considered an understatement of epic proportions, but Emotion is a form of intelligence that goes well beyond any traditional connotation of the word. **Emotion** may be equated with feeling or desire. When emotion is positive and healthy, it motivates us to achieve an objective and develop; when it is negative and unhealthy, it often results in failure and restricts our development.

Admittedly, unbalanced emotion may sometimes have a negative impact on any other intelligence we have listed thus far. Tent revival enthusiasm may lead one to a blind adherence to unexamined beliefs. Raw, unchecked emotion may also produce shallow assessments, incorrect procedures, or skewed priorities. Emotionalism, such as rage or humiliation, may even result in physical harm, as evidenced by the stereotypical (but very real) "crime of passion." In short, if not used objectively, emotion can cloud our judgment and cause erratic behavior.

We should realize that this intelligence, when balanced, carries very positive connotations. We should not lose sight that motivation and desire have resulted in astonishing achievements throughout the history of mankind. The movement of man from small bands of hunter-gatherers to large populations that use law, reason, and technology represents an evolutionary drive toward potential that is nothing short of astonishing, even when allowing for man's inherent imperfections. The intelligence of Emotion has driven man, individually and as a species, toward new levels of development. From fire and primitive tools, man has advanced to making computers and satellites because of the desire to know and explore. Furthermore, if we take a cross-section of any period in recent history, we see how Emotion translates potential into positive developments. Man has produced great works of art in the areas of music, literature, and painting. Without consummate motivation, it is doubtful that Michelangelo could have lain on his back high atop scaffolding in the Sistine Chapel for the better part of four years while painting the fresco depicting the creation of man. There is an old adage that says, "Man must suffer for his art," and if this is true, it is surely the desire to create something transcendent that enables him to endure hardship, poverty, or criticism as he attempts to give artistic shape to his inner vision, transforming the abstract to the concrete.

Emotion also demonstrates man's compassion and virtue. Jonas Salk worked tirelessly to develop a vaccine for polio, testing it on himself before attempting to introduce the refined product to the medical and pharmaceutical communities. The result was a virtual eradication of polio during the twentieth century. With the same drive to help mankind, Mother Teresa cared for the sick and dying of Calcutta with a selflessness that could only have become fully realized through the intelligent functioning of Emotion. One need not be aligned with any religious doctrine to appreciate the high standard of virtue that her Emotional intelligence produced.

We can also see desire as a driving force in the exploration of space. In the early 1960s, President Kennedy spoke of space as the new frontier and challenged the nation to send a man to the moon and return him safely by the end of that decade. While this goal was largely inspired by a desire to beat the Soviets in a space race, the scientific and engineering skills involved in meeting this challenge in so short a time might be regarded as miraculous but for the fact that we can say in hindsight that NASA and various contractors were able to do two things: first, they used incredible focus on the tasks at hand; and second, they approached these tasks with an ambition born from intense desire. Without supreme desire, it is doubtful that scientists and engineers could have overcome the thousands of obstacles they encountered in meeting President Kennedy's challenge.

The role of Emotion in attaining goals can also be very powerful when coupled with Physical intelligence. Consider the mother who is suddenly able to lift a massive weight, such as an automobile, that is crushing her child. It is strong emotion, summoned almost instantaneously, that chemically produces superhuman strength in her muscles.

To say that emotion clouds reason can be shortsighted. On the contrary, emotion is simply one possible avenue that reason can use to express itself. As is said in football, "On any given day, one team can beat any other team in the league." How? By being emotionally charged. Some examples of Emotional intelligence are being cool under fire, motivational speakers, acts of inspiration, driving to achieve, passion, desire, etc.

A Not-So-Trivial Pursuit: The *Mental* Functional Intelligence

The next functional intelligence is the **Mental**. This refers to knowledge and IQ, as well as the ability to use reasoning and critical thinking skills. To use a colloquialism, it is the use of brainpower.

When we discussed education, we noted that application of knowledge is a more desirable goal in the long run than the simple storage of facts. This does not at all imply, however, that memorizing is a bad thing. This is one more example of how NATI seeks not to judge, but to balance. Gaining knowledge is highly desirable since the more facts we have, the more variables are at our disposal when we need to engage in higher forms of reasoning or apply other kinds of intelligence. A general on the battlefield cannot assess where to deploy his troops unless he has been given facts upon which to base his decisions. Procedures to send man to the moon hinged on the scientific knowledge possessed by thousands of engineers, knowledge attained collectively through millions of hours of study and training. Even law enforcement is contingent upon a police officer's knowledge of the statutes in his jurisdiction.

Research now indicates that acquiring knowledge appears to help prevent Alzheimer's disease. Each time we learn a fact, it is stored in our brains, first in short-term memory and then, if reinforced often enough, in long-term memory. It appears that doing this repeatedly can actually create new neural pathways within the brain (or, at the very least, reinforce old ones). In fact, atrophy is less likely to occur in the first place when people are encouraged to read, play games, solve riddles, and memorize facts. Using our Mental intelligence is a small price to pay to keep this insidious disease at bay.

Many heroes from film and literature have been grounded in Mental intelligence. Sherlock Holmes, the creation of English writer Sir Arthur Conan Doyle, was able to solve crimes by using a highly refined mental capacity, known in the genre of mystery as ratiocination (or deductive reasoning). To be sure, Holmes' intelligences were razor-sharp, enabling him to amass more information than the average person, but the process by which he solved his mysteries was primarily cerebral. His mental powers and sense of logic might well be said to rival the most sophisticated analytical computers in existence today. (The human brain, of course, is a computer, and it is the contention of NATI that everyone can enhance the Mental functioning of his or her "biological PC.")

The legendary Mr. Spock from the *Star Trek* television series is another example of the premium we have placed on Mental intelligence. The character of Mr. Spock came from the planet Vulcan, a planet whose race had shunned emotion in order to attain a strictly rational, intellectual approach to life. In weekly episodes, Mr. Spock was able to solve complex problems in a way that made logic and mental intelligence appealing to the viewing public. The fact that this fictional character attained cult status indicates that people realized, even if only on an unconscious level, that mental functioning was an important part of their overall make-up, one that they wanted to develop more.

Other examples are numerous: grand masters of chess can think dozens of moves ahead during the course of a game; child geniuses sometimes start college before adolescence; savants can perform mathematical computations in seconds or play a complex melody on the piano after hearing it only once.

Further examples of people who use Mental intelligence are educators, scientists, accountants, scholars, intelligence agents, analysts, etc.

Brainwaves And Beyond: The *Intuitive/Spiritual* Functional Intelligence

The fourth and last intelligence in the Functional Group is the **Intuitive**, or **Spiritual**. Let me say at the outset that the term "spiritual," as used in NATI, does not refer to any kind of religious practice. Associations between overtly religious people and spiritual energy are sometimes made because the practitioners of certain religions seem to have a strongly developed psychic sense. Numerous surveys, however, reveal that people from all areas of life report intuitive phenomena.

To function intuitively is to tap into a higher force or energy that elevates a person's consciousness to a different level of functioning. Such functioning has traditionally been regarded as the sole province of mystics and seers, but this limited view is being challenged more and more by ordinary people exploring the development of their spiritual, intuitive sides. One such person is biologist Rupert Sheldrake. In his book *Seven Experiments Which Could Change the World*, he explains that spiritual growth can be attained by learning to recognize synchronicities.[2]

The term synchronicity, which was coined by Jung, refers to a meaningful coincidence, one that could not be expected to happen under circumstances governed by mathematical probability. Sheldrake regards synchronicity as a natural occurrence in what both physics and mysticism now call a "responsive universe." Moreover, Sheldrake believes that ordinary people can become more aware of synchronicities in their lives by paying closer attention to small events, such as a chance encounter with a friend, an unexpected phone call, the song playing on the car radio—literally anything that comes into their field of awareness. Sheldrake believes that when people pay close enough attention, they begin to notice patterns emerging that represent the intelligence of the universe, dispersed in the non-local manner we have already alluded to, attempting to communicate with them.

Despite a greater interest in metaphysical literature in the past several years, there are still skeptics who claim that such occurrences of synchronicity are no more than people seeing what they want to see. The idea is really not radical at all, however, if we accept the NATI premise that intelligence is uniformly present in all of creation. If reality itself is intelligent, and we, who are intelligent beings, are a part of reality, then it is quite normal to be in connection with the Whole of which we are Parts. That we have access to universal intelligence and can have free and easy discourse with a larger field of consciousness is therefore not at all unusual, but rather a normal condition.

There is actually a scientific basis for this idea of access to a non-local intelligence, but to understand it fully we need to touch on the areas of brain chemistry and holography. Glial cells are distributed throughout our brain in random, three-dimensional patterns. The work of Renato Nobili, an Italian physicist, demonstrated that sodium and potassium ions are able to move through glial cells, forming oscillating wave patterns that appear identical in frequency to "Schrödinger waves."[3] (Erwin Schrödinger was the first scientist to express in precise mathematical calculations the nature and movement of the quantum wave.) The ramifications of Nobili's work are wide-ranging, for we now have evidence that quantum probability waves are part of normal brain function. This leads directly to the theory of brain holography.

We will investigate holography at great length in a later chapter, but for now, it is sufficient to say that a hologram is the product of a superposition of images. It is this superposition, or layering of light waves, that yields the three-dimensional holographic effect we see in pictures that appear to have a glazed appearance. What Nobili maintains is that the same superposition occurs in the brain because of the three-dimensional distribution of the glial cells in the human cortex. The sodium-potassium wave activity within these cells is literally capable of forming holograms within the brain. This tallies completely with medical research in the field of Alzheimer's regarding the distribution of memory throughout the cortex. Fred Alan Wolf, physicist and author of numerous books on quantum theory and the nature of reality, says in *The Dreaming Universe*,

> When we put all of this together, we have a reasonably good expectation that the brain can act as a holographic medium and that a vast amount of memory exists via slight changes in the glial cells, which act as absorbers of Schrödinger wave energy. Not only this, but because of the superposition principle, glial cells can absorb superpositions of wave information, some coming from recent events and some coming from past events. Depending on the reference wave that excites the glial cells, associated memories can be evoked.[4]

Einstein's brain, when autopsied, was found to have a higher than normal concentration of glial cells in his visual cortex, leading many to believe that this was partly responsible for his ability to visualize so many abstract concepts.

The holographic brain, therefore, is our interface with nonlocal (universal) intelligence. The brain is able to see a "whole picture" of reality by virtue of its holographic nature. As Wolf notes above, the brain is not limited by space or time and may receive messages from both the past and future. (This is a corollary of quantum theory, and the interested reader will find more on the subject in Wolf's *Parallel Universes.*) The idea that humans can be in touch with all things and all times is one of the foundations of Buddhism, Hinduism, and Taoism.

How we use such whole brain thinking in our own lives is, of course, up to us. NATI does not prescribe any particular goal, but Intuitive intelligence has many applications. People who invest in the stock market often balance their economic forecasting skills with old-fashioned intuition. Likewise, entrepreneurs often "go with their gut" as much as they use knowledge of business when it comes to decision-making. Professional gamblers are usually armed with a great deal of information pertaining to gaming rules and odds, but they also use intuition to know when taking risks might be beneficial.

There are numerous other examples of Intuitive intelligence at work, such as knowing when the phone is about to ring or who will be calling. Many people claim to be able to sense future happenings, such as earthquakes, kidnappings, assassinations, economic recessions, and numerous other events. During the cold war, the governments of both the United States and the Soviet Union employed people with intuitive abilities to participate in experiments such as "remote viewing."

Perhaps the most famous examples of intuition are those instances when a parent knows that his or her child is in imminent danger and is moved to take sudden, life-saving action (or at times to simply sense that a child is encountering some kind of difficulty).

All of these examples attest to the reality of Intuitive intelligence.

Finding Functionality

Let's take a hypothetical look at applying the four functional methodologies for the sake of comprehension.

Assume we have a high school classroom with typical, average students. The issue is that the class, both individuals and the class as a whole, is not doing well! The objective is to enhance their overall and individual performances. Let's first look at the physical/material intelligence approach utilizing a polar scheme.

On the negative side, we threaten the students. "No sports," says the teacher. Painfully, little happens. On the positive side, we bribe them. "No homework on weekends," we say. Still, there is little effect.

So now we try applying the mental approach. We bring in the president of Mensa to explain the pros and cons of the situation, such as not learning enough to provide the students a good lifestyle. Alas, nobody understands the speaker's vocabulary! Not even the teachers! Again, no results.

Then we try the emotional approach. In comes Tony Robbins, the famous motivational speaker and ye gads—it works! Up go the scores, as well as the class's attitude! But, after five days things are as they were before. Ye gads again! We call Tony again, but he's in Hawaii doing a fire walk, so he's not available. We get mad at the students. That doesn't work! We preach to them about their lack of will. Nothing works. Ye gads still again!

Our last hope is the spiritual intelligence. Oh, no! Are you going to have a séance? A revival meeting? It's still ye gads! But no—that's not the case. Spirit is something else. Sports bookmakers give the football home field advantage a three-point spread. That's because the crowd generates tremendous support for their team—a spirit, if you will.

The same thing holds true for team spirit. Now we're onto something. How about a classroom spirit—a class culture, an image for all to achieve? So we call Joe Torre, manager of the New York Yankees. The Pride of the Yankees is a universally recognized Yankee spirit. Joe talks to the kids about being part of a successful team, part of a whole. He emphasizes the significance of helping yourself and the team by reaching for ever higher standards, developing greater potential, the coolness of being an achiever in a group of achievers, no matter how small one's contribution is. He shows the kids that this is achievable but that failure may still occur. Why? Because anything else is unrealistic and that failure and weakness are our paths to developing Potential. This frees us to be honest with ourselves and others.

Triple ye gads! It works! Kids start to help each other, try harder to be part of this special unit, and develop pride and recognition in being a member of this outstanding group. I think you get the point!

Summary

The Functional Group of intelligences corresponds to the four spiral aspects of cell development. Every kind of function that people can perform is carried out physically, emotionally, mentally, or intuitively.

Physical intelligence is literally the manifestation of potential energy and may be regarded as the intelligence in all matter. In humans, this intelligence is manifested in physical bodies and the activities they engage in.

Emotional intelligence may be related to feelings, desire, or motivation to achieve an objective. While we often equate emotion with impulsivity or rash action, it is capable of providing the impetus to reach even hard-to-attain goals.

Mental intelligence involves the use of reasoning and critical thinking skills. It also includes the acquisition of knowledge and how it is applied in various situations. It may be thought of colloquially as brainpower.

The last intelligence in this group is the Intuitive. This intelligence is akin to moving past pure reason and "playing the hunch." It involves sensing or inner revelation. In other contexts, it is related to becoming attuned to synchronicities, or meaningful coincidences. As Socrates declared, intuition is pure reason.

PART THREE

POLARITY
THE INNATE ACTIVATOR

Chapter Eight

Polarity

Without Contraries there is no progression.
Attraction and Repulsion, Reason and Energy,
Love and Hate, are necessary to Human existence.
*—William Blake, in "**The Marriage of Heaven and Hell**"*

Long and short complete one another.
High and low rely on each other.
Pitch and tone make harmony together.
Beginning and ending follow one another.
—Tao 2

At this point, we move into a part of our new system that is vital, innate and quite mischaracterized. Across an entire cultural spectrum, the Western mindset frames issues in terms of yes/no, guilty/innocent, right/wrong and so forth. As we saw in earlier, there is a drive toward individual accomplishment in the West, although trying to obtain the largest amount of money in the shortest amount of time doesn't leave much room for nuance. Western society seeks instant gratification and perfect justice. We want food handed to us through drive-up windows; we want movies made available "on demand" on our cable TV systems; and we want the good guy in the white hat to triumph in each and every conflict.

In our haste to get results, we see life in simplistic terms since polarities can be recognized without a great deal of thinking. Notions of good and bad are visceral and can produce the Great Restrictors of Fear, Ego, Ignorance, and Self-deception. When we think with only our emotions, we are usually driven to polarities.

Consider the bumper sticker that was so popular during the Vietnam era: AMERICA: LOVE IT OR LEAVE IT! One may argue that this bit of raw emotion hurled at protesters didn't leave much room for the kind of discussion that our democracy was founded on. To some, the visceral appeal to patriotism ignored the fundamental American rights of free expression. People opposed to protest would scream at demonstrators that U.S. troops in Asia were fighting for the right of protesters to march in the streets ... so the protesters should therefore go home and support the boys in Vietnam. The contradiction in logic was not perceived. Ironically, the protesters often behaved no better, but instead would hurl insults at those criticizing them, causing shouting matches that ended only when the police dragged both parties away.

These sentiments are unfortunately still very much with us. For any given cause, we are told that, "If you're not for us, you're against us." For/against, either/or—these are the same extremes that we have clung to since childhood. Pro-choice activists insult those who are adamantly opposed to abortion, while pro-life activists attack opponents to proclaim the sanctity of life. Ignorance reigns supreme.

The list goes on and on: gay marriage, sex education, evolution, healthcare, foreign policy, tax cuts, and euthanasia are but a few of the inflammatory issues that divide the United States. We are culturally conditioned to approach issues in terms of right and wrong, with virtually no willingness to concede that some answers might lie in the proverbial gray areas of life.

Is it any wonder that, having witnessed accusations and criticisms grounded in polarities from first grade through the twelfth, many people suffer from low self-esteem and depression? If we don't have the right kind of body, mate, house, car, and job, the resulting perception is that of failure.

OLD DEBATES

Even science is not immune from polarized thinking. One of the oldest controversies in psychology is whether nature (genetics) is more important than nurture (environmental factors) when it comes to explaining a person's behavior. Do we inherit behaviors the way we inherit eye color, height, or build? Or is the way we act more a product of our environment and upbringing? Can a stable, loving home override unhealthy tendencies? Can good genes lift us above a less-than-ideal home life? Both nature and nurture would seem to be important factors in everyone's development, and yet the debate continues as psychologists and geneticists square off in professional journals, citing their latest research that allegedly proves love conquers all or criminals are made in the womb.

Far from finding any common ground, researchers seem to become more and more polarized. Some medical studies claim that there are genes for obesity, homosexuality, and criminal behavior. Meanwhile, psychologists continue to publish findings showing how twins separated at birth can turn out differently if adopted by separate families and raised in different environments. The truth is that genetics and environment affect each other in many complex ways; trying to separate the influence of each is as impossible as "unbaking" a cake to identify its ingredients. Trying to affirm only one position in this debate is tantamount to saying that a coin has only a head or a tail.

It's as if we're hardwired by culture to carry prejudice and shallow assessments from the schoolyard into serious areas of scientific investigation, where opinions have no place. Our focus on extremes not only causes arguments and frayed nerves, but it hampers humankind's ability to study both itself and its environment in order to achieve what NATI regards as the paramount mission in life: the development of our potential.

Even brilliant physicists have reacted to opposing theories with emotion rather than reason. The theory of wave-particle duality, perhaps the greatest polarity in all of nature, caused considerable consternation among many theoreticians. Einstein himself was profoundly unsettled by the entire notion of uncertainty and for the most part turned his back on quantum mechanics as he sought a way to unite all forces in nature that would represent a Grand Unification Theory. He was convinced that quantum fluctuations could be predicted if he could find "the bigger picture." The uncertainty inherent in wave-particle duality was not something he could accept as a final principle. Though Einstein could not refute the mathematical data behind Bohr's Copenhagen Interpretation, his visceral response to the Uncertainty Principle was quite human, his genius notwithstanding.[1] He thought that Bohr's quantum theories represented, at best, a small piece of a larger puzzle.

There's No Denying

Polarities are an integral part of the universe. No one can deny that individuals display a wide range of behaviors. People can be passive or aggressive, weak or strong, open-minded or closed, introverted or extraverted, abstract or concrete, optimistic or pessimistic, objective or subjective . . . and these are only a few of the more recognizable characteristics that can be observed in humans. Polarities simply exist!

The idea of opposites is even built into the fabric of subatomic particles. Antimatter is not just an invention of science fiction, for electrons have positively charged companions called positrons. Similarly, protons and neutrons are composed of quarks, small bits of matter that possess different qualities of spin, called *up* and *down*, *top* and *bottom.* Polarity seems to be responsible for keeping things together in a most fundamental way in our universe. Indeed, there are dozens of subatomic particles, such as gluons and muons, that act as glue for the strong nuclear force within the atom. We know, however, that reality somehow uses these opposite particles to produce the observable universe. There is a tension and dynamism among these elemental bits of matter that produce the paper you are now looking at or the furniture in your room. Polarity allows the potential of the universe to manifest itself. *The cosmos regards tension between polarities as a creative dynamic, and one of our tasks is to find some way to use this tension instead of always shunning or avoiding it.*

A more productive outlook on the energy inherent in any system of opposites can be seen by looking at the work of William Blake, an English poet of the early Romantic period. Blake felt that "in political history, in Christian theology, in great works of literature such as *Paradise Lost*, and finally in the nature of each individual, diabolical energy and restrictive convention, sin and virtue, chaos and order, innocence and experience, are in eternal combat with one another." Blake's vision of Polarity is seen most clearly in his work *The Songs of Innocence and Experience.*[1] Originally published as two separate volumes, one illustrating innocence and the other illustrating experience, the poems were published as a unit in 1794, with each poem on innocence corresponding to a companion poem on experience. The poems intentionally represent literary and philosophical polarities. The innocence of a lamb, for example, is contrasted with the savagery of a tiger in "The Lamb" and "The Tiger," respectively. Blake, an avowed skeptic in all religious matters, intentionally challenged his reader to consider that the same force, both divine and diabolical in nature, was responsible for the creation of the lamb and the tiger. Needless to say, some of Blake's contemporaries considered the comparison of the Lamb of God to a ferocious tiger to be more than a little blasphemous.

While Blake believed that life was a dynamic process, he also believed it could be static at any given moment depending upon an individual's perception or focus. There is a natural correspondence between this idea and Heisenberg's Uncertainty Principle. Translated into modern scientific terms, Blake was saying that we choose to observe events in space-time as either darkness and sin or virtue and light. To Blake, such simplistic observations could not account for the tension and conflict that permeate life.

Blake's philosophy of dynamism and opposites is reflected perfectly in the writings and theories of Abraham Maslow.

Maslow saw that the arbitrary pitting of good against evil resulted in the possibility of Unrealistic Perfectionism. He believed that the historical search for a utopian society produced simplistic notions such as:

- Let us all love one another.
- Let everyone share equally.
- No one should have power over anyone else.
- The application of force is always evil.
- People are never bad—only unloved.

These Utopian ideals are exactly what Blake was writing about in "The Lamb," ideals that he vehemently rejected in "The Tiger" and other songs of experience as being unrealistic and naive. Like Blake, Maslow believed that such a one-sided approach to life could not account for the full spectrum of human behavior. In Maslow's estimation, expectations of perfection led to disillusionment and apathy.

War and Peace

When I first set out on my search for truth many years ago, one of the first things I sought was a concept that would include opposites without trying to negate or deny them in any way. My journey led me to Taoism and the Eastern belief in *yin* and *yang*, opposing forces representing female and male, passivity and aggression, negative and positive. In Eastern thought, *yin* and *yang* are seen as complementary parts of a whole, not adversaries striving to triumph over one another. The *Tao* advocates a balance between forces, not antagonism. In *The Tao of Inner Peace*, Diane Dreher writes,

> When we're caught up in dualism, conflict invariably turns into combat. Fearful and defensive, we project our negative shadows upon our opponent, whom we see as the cause of all our problems. We define conflict resolution as a matter of either "winning" or "losing," defeat-

> ing our opponents or being defeated. All other options vanish, and instead of using our energies to solve problems, we turn in fury upon our perceived enemy."[3]

The same prescription for living was given by the Buddha in his Fourth Noble Truth, in which he advocates the Middle Way, a path that avoids extremes. For the Buddha, it was this "third way" that ultimately has the ability to release mankind from *samsara*, the ceaseless cycle of birth and death driven by karma, the never-ending chain of cause and effect.

I can unequivocally say, therefore, that negativity is something we should learn more about as we attempt to further our understanding and personal development. Indeed, there is stability in all of nature based entirely on this union of opposites, including good and evil. Without a doubt, human existence itself is modeled on the tension between extremes, for life is always placing us at the crossroads of Polarity. Whenever we employ a plan or engage in an action, we are choosing against its opposite. Life constantly pushes us to make choices so that we may evolve and grow while contending with alternate ideas and their consequences. Sometimes, evil is inextricably bound up in our decision-making processes and we need to come to terms with this fact.

For instance, we can view war as dire tragedy with no redeeming value, and yet the total absence of war might be equally tragic. The Civil War was the bloodiest conflict our nation has ever seen, with a combined total for both North and South of over 600,000 casualties. Abraham Lincoln was faced with excruciatingly difficult choices, but our society today now espouses rights and laws that might not exist if the Civil War had been averted and the South had developed a legal code of slavery in the twentieth century. The road to racial equality was one that ran through the very heart of conflict.

Harry Truman is criticized by many for dropping the atomic bomb on Hiroshima and Nagasaki. Whether or not World War II could have reached a successful conclusion in the Pacific theater without the detonation of nuclear weapons is a matter of debate, but one thing we do know for sure is that the possibility of total nuclear annihilation resulted in a middle course between the United States and the Soviet Union in the decades after the Second World War. Both sides practiced the principle of deterrence as a way of avoiding all-out nuclear conflict. There was peace, but it was an uneasy peace with missile silos housing thousands of nuclear warheads ready for launch. If Truman had not made the decision to drop the two atomic bombs, would mankind have refrained from using more advanced nuclear weapons in later years? It is impossible to tell, but the wars mentioned above show that tension yielded long-term compromises representing workable solutions. The issues of civil rights and global peace are examples where a third way evolved from the darkness that is war. (I am not condoning war, but merely using it as an example.)

Evolution reveals the same pattern. The history of all life on earth shows that species survival depended on strength, power, predation, and domination. Once these traits had proven successful, species were able to find a stable environment in which to propagate. The often violent mechanism of natural selection resulted in stable, harmonious ecosystems.

On a much smaller scale, we can see the dynamics of Polarity in an issue that was discussed in our explanation of Order and Process, namely predictability vs. variation. Taken together, these traits represent the kind of polar characteristics we are likely to encounter in our everyday lives. Some people are creatures of habit; others are more inclined toward constant exploration and change. These behaviors are neither good nor bad. What is important, therefore, is how we deal with them. Once again, the example of Japanese manufacturers is germane. Japanese executives structure time for exercise and recreation during normal working hours, satisfying the need for both predictability and variation. In the corporate world, the area of conflict resolution depends on getting parties with opposing viewpoints to accept a third alternative. If managers at the International Widget plant are unhappy with the low output of their workers, they may become hardnosed and confrontational. The workers, discontent over poor lighting and heating conditions, may threaten to strike. A skilled conflict resolution specialist will explain to management that you can catch more flies with honey than vinegar. In this case, the honey is adequate light and heating. The specialist will then explain to employees that productivity is the key to a successful business and that unless more widgets are made, management won't have the necessary money in its payroll account to cut the workers' checks. The solution to the problem lies in a compromise that acknowledges the importance of both widgets and comfortable working conditions.

This concept of a middle way can be seen in relation to any NATI intelligence. My own marriage is a good example. I am a Process-oriented person, while my wife Angela is geared toward Details. This is a perfect scenario for conflict (which was certainly present in our pre-NATI days). I now use her inclination for Details to augment my functioning, while she uses my Process orientation to complement her own actions (when she feels like it).

By transcending the extremes of any two polarities, a new reality is formed. This transcendence is the essence of dynamic development and the very heart of realizing our potential! Everyone in the world possesses the thirteen intelligences and can be considered an individual matrix that functions in conjunction with the energy of polarities. No two matrices are identical, and any individual matrix may be regarded as an open system capable of adaptation and change. Fundamental to the NATI program is that we all have strengths and weaknesses. As in the example of my marriage, *we can use Mirroring to see our weaknesses reflected to us by others and do something about them!*

When we become aware of our thirteen natural intelligences and accept the existence of polarities without judging them, we can begin to implement changes in our lives. Through awareness and knowledge, we can travel the road to self-actualization and peak experiences. Winning, losing, and enemies—these concepts take on less stature when we employ our innate abilities to solve problems.

Life and Death and Choice!

From the standpoint of our everyday lives, nowhere do we see issues of Polarity more clearly than in how we approach health-care and the management of disease. When we face choices of comfort vs. suffering, disease vs. wellness, life vs. death, we are, of course, confronting polarities. Given any ailment, serious or otherwise, we can either ignore it or go through every test known to medicine in order to find a cure. Granted, there may indeed be times when ignoring an ailment is in our best interests. Holistic practitioners often tell their patients to refocus awareness from the process of illness to the process of living. Conversely, there are some diseases that call for numerous tests to be performed before a physician can make an adequate diagnosis. More often than not, however, there are choices to be made.

Furthermore, we can see how all thirteen intelligences operate together by considering how a medical problem is handled from the moment of detection to the time it is cured. Awareness is used to gain focus on an ailment. Tests and assessments are made to make a diagnosis and prescribe procedures for getting better. What we believe about the diagnosis will determine how we physically and verbally express ourselves and will furthermore determine how motivated we are to get better. For instance, if we believe in the medical models we are presented with, we will have a greater desire to engage in physical activities or mental thought patterns that can use the energy of polarities to set the healing process in motion. Later, when we reach the stage where we can focus more on living than being ill, we are essentially placing the diseased part of the body back into the context of a complete, functioning organism. Finally, when we feel better, we are happier and better equipped to handle life's challenges, enabling us to move on.

By now, I hope you are able to see three very important NATI truths. First, it is literally impossible to talk about life without naturally invoking the thirteen intelligences. Second, our thoughts determine our actions at every step. Third, we are dynamic systems (or matrices) comprised of intelligence and polarity that can be used to actualize our potential in any area of life.

Transcendence

Because the NATI structure follows specific laws of nature (i.e. cellular development), it is validated many times over, as well as by its demonstration of consistency, accuracy, and reliability. To put the final bit of polish on the concept of opposites, we will conclude this chapter by seeing how quantum physics deals with the issue of Polarity.

We know that Niels Bohr and Werner Heisenberg demonstrated the wave-particle duality of light and matter. This duality was termed the Theory of Complementarity by Bohr. Bohr held to his idea of complementarity in the face of criticism from the likes of Einstein, Schrödinger, and Planck. In response to his critics, Bohr himself admitted that picturing matters of quantum probability was a bit dizzying, but he steadfastly maintained that the microworld of the electron need not be visualized in the macroworld. Bohr had a fundamental grasp that complementary aspects of reality didn't have to be seen to be accepted. These aspects—these polarities—simply existed.

Regardless of experimental factors, therefore, electrons transcend both states. The mental ability to accept wave-particle duality is itself a transforming act, a creation of a third path of thinking that rejects the either-or mentality driving so much of Western civilization. It is a movement toward oneness and cultural integration. It is a quantum leap that gives us the freedom to perceive alternatives as we direct our focus more and more to the abstract realm where matter and energy are unified. All ideas, actions, and matter originate from potential energy—the living, pregnant void.

String theory may yet give us a more concrete way to visualize this duality by proving that all forces in nature are united by vibrating loops of energy. In the meantime, NATI provides everything we need to know about unity in a single word: complementarity.

Further Polar Characteristics: Personality Orientations

We have talked a great deal about the transcendence made possible by complementarity in the last few pages, but I want to emphasize that NATI does not require anyone to adopt specific religious or cultural beliefs that are sometimes connoted by this term. While using hard science and respected philosophical traditions to explain various concepts, we are ultimately concerned with helping people do things more smoothly and with minimum stress, hardship, and worry. The transcendence we strive for is concerned with accountability, responsibility, understanding, and development, not with doctrinal beliefs.

To this end, look at the list below, which represents some of the many **Personality Orientations** that represent polarities in our everyday activities and relationships.

potential	=	possibility
actualized	=	done or accomplished
introverted	=	inward-oriented
extrovert	=	outgoing
open	=	willing to accept, expansion
closed	=	unwilling to accept, restricted
objective	=	impersonal position
subjective	=	personal position
variant	=	focus on dynamics, change, and flexibility
invariant	=	focus on fixed and static patterns

self	=	self-centered
selfless	=	other-centered
rote	=	doing things automatically
thinking	=	examining objects and ideas carefully
abstract	=	relative, archetypal
concrete	=	factual, tangible
positive	=	plus
negative	=	minus
defensive	=	guarded
aggressive	=	forward
weak	=	lacking
strong	=	capable
probable	=	likely to occur
determinate	=	definite

Everyone is an individual matrix with strengths and weaknesses. By identifying and addressing these weaknesses, you can reach virtually any goal you choose. When you bring focus to bear on weakness, you can, among other things, determine what NATI calls your Current Operating Procedure, or COP. COPs are mostly unhealthy behaviors that corrupt everyone's naturally clear and efficient matrix at some point by rigid, institutional thinking. By becoming aware of your COP, you can learn how the Great Restrictors hold you back to some degree or other. It makes no difference whether you want to improve your golf game, quit smoking, be a better student, or enjoy healthier personal relationships. Identifying your COP can help improve your life.

Always remember that we should never be afraid of weakness. You may therefore recognize traits in the above list, aspects of yourself that you would like to change. You may be afraid to try new things and listen to the views of others. You may be introverted and passive, fearful of expressing how you really feel.

Perhaps you are weak and defensive in certain situations, always on guard against what you perceive as possible harm. Maybe you're not aggressive enough on the golf course. If you're trying to quit smoking, you may feel that the chances are stacked against you, so that your attitude is "probable" rather than "definite." Regardless of what weaknesses you may have, you can use them to grow by first becoming aware of them and then examining your beliefs about them, such as:

- Am I afraid to listen to others because they may think they are smarter than I?
- Do I withhold my point of view for fear of being ridiculed?
- Do I think I will be hurt in a relationship because my previous partner treated me with cruelty?
- Am I worried about what others are thinking when I swing a golf club?
- Do I think any attempt to stop smoking is doomed because I tried before and failed?

Next, we will examine the theory of systems in order to understand our matrices even better. When we are able to appreciate how systems work—when we understand how energy and information keep a system open and fluid—we can take the quantum leap toward holistic, natural thinking.

Summary

As wave-particle duality implies, polarities are an integral part of the universe and keep things together in a fundamental way from the standpoint of quantum mechanics and particle physics. We should, therefore, never regard opposites as bad. By transcending the extremes of any two polarities, a new reality is formed according to the Principle of Complementarity. We can use the tension and energy of opposites to create what Buddhist philosophy calls the Middle Way.

PART 4
SYSTEMS

CHAPTER NINE

THE PRINCIPLE OF SYSTEMS

In order for us to properly implement our principles of nature and innate intelligences, we need to understand the way in which those factors operate.

Systems are a part of our lives...and our lives are parts of systems, but we are scarcely aware of this fact. This is because we don't pay much attention to certain phrases we hear over and over again. We constantly live and interact within ecosystems, educational systems, weather systems, and systems of government. We use healthcare systems, communications systems, computers systems, banking systems, and transportation systems. These are just a few of the systems we use on a regular basis, for all life on earth can be organized into social, technological, or biological systems. The point here is that we focus on the *type* of system we interact with, but not with the actual structure of a system. We take for granted assorted and questionable components of given systems, including the symbols used to communicate ideas, some of which are highly complex and sophisticated.

But what makes a system a "system"? Is "system" just a name for a group of people or objects, or does the term imply something more specific, such as a network of relationships? It's important to answer these questions, for we have already alluded to Natural Thinking & Intelligence as a system many times in the course of these pages.

What is a System?

The following characteristics were gleaned from *The Art of Systems Thinking* by J.S. O'Connor and Ian McDermott. They define the most important aspects of any system as follows:[1]

- A system is an entity that maintains its existence and functions as a whole through the interaction of its parts. The behavior of different systems depends on how the parts are related rather than on the parts themselves. Therefore you can understand many different systems using the same principles.
- Systems form part of larger subsystems and are composed in turn of smaller systems.
- The properties of a system are the properties of the whole. None of the parts has them. The more complex the system, the more unpredictable are the properties of the whole system. These whole system properties are called *emergent properties.* They emerge when the whole system is working.
- Breaking a whole into its parts is analysis which helps one to gain knowledge. Building parts into wholes is synthesis. When you take a system apart and analyze it, it loses its properties. To understand systems you need to look at them as wholes.
- "Detail" complexity means there are a great number of different parts.
- "Dynamic" complexity means there are a great number of possible connections between the parts because each part may have a number of different states.
- Each part of a system may influence the whole system.
- When you change one element, there are always side effects.
- Systems resist change because the parts are connected.

A Brief, Painless History of Systems Theory

The theory of systems can be traced largely to one man: biologist Ludwig von Bertalanffy, born near Vienna in 1901. In the 1930s, von Bertalanffy formulated Organismic Systems Theory, an approach to systems that, quite simply, said that a dynamic process existed within organic systems. The theory essentially focused on biological organisms in which metabolism strived for growth and equilibrium. Anyone the least bit familiar with high school biology will recognize that this describes cells, plants, animals, and humans.

In the 1940s, von Bertalanffy became interested in thermodynamics, or the transfer of heat into other forms of energy. Like his colleague Ilya Prigogine, von Bertalanffy was intrigued by the fact that the Second Law of Thermodynamics, which states that everything moves toward stasis and entropy, did not apply to all systems. Von Bertalanffy therefore formulated his Theory of Open Systems, showing that energy flowing into a system can bring it to a steady state (or a condition of self-regulation).* If open systems did not exist, everything in the universe would soon run down, dissipating into chaos. Each morning, though, millions of people get out of bed and become part of the many systems mentioned in the opening paragraph of this chapter. Open systems are all around us and import energy to maintain themselves.

In the formal scientific study of systems in the 1940s, the terminology of biology and physics was used by von Bertalanffy and his colleagues. Von Bertalanffy wanted a more accessible and comprehensive approach to systems, however, and so introduced General Systems Theory. GST, as it is known, is a way of describing virtually *any* system in ordinary, non-formal language. Thanks to von Bertalanffy's simplified approach to systems, we can list the two most important principles of General Systems Theory:

---

*In effect, it becomes self-reliant, consistent, stable.

- Individuals and societies can be open or closed models. A closed system possesses components that interact only among themselves and not with the environment. This type of model is not a growing, evolving system. This means being internally oriented and disconnected from the environment. In behavioral terms it represents closed-mindedness and linear thinking.
- An open system is comprised of components that can interact with each other after receiving an input of matter, energy, and information from the environment. The notion here is connectedness, open-mindedness, and integrated thinking.

According to von Bertalanffy, a system is not a "thing" so much as a dynamic pattern of organization in which the system as a whole is more important than its parts by virtue of the constant interaction of its components.

This is the essence of General Systems Theory, which has become a paradigm for understanding the dynamics of all social, technological, and biological systems since the 1950s. Part of von Bertalanffy's genius was to take a complicated set of variables and explain them in simple language that is universally applicable to all systems.

In keeping with our ongoing comparison of science and Eastern thought, we can say that open systems are philosophically analogous to the beliefs of Buddhism, in which physical objects are transitory. Buddhism is "event-oriented," especially Zen Buddhism, which emphasizes participation in life's activities. Dynamic interaction with the environment is the preferred method of "being" to disciples of Zen.

Open and Closed for Business: Open and Closed Systems and Chaos Theory

The average home heating unit is a closed system. Water is heated and sent throughout the house via a network of pipes. When a certain temperature is reached, the thermostat turns off the heater. It is an efficient system, but it is really nothing more than a feedback loop and is closed to outside information. This is true for all cybernetic systems, which are predictable because they regulate themselves within strict limits prescribed by their purposes and structures. Von Bertalanffy believed that cybernetic models used circular causality and were therefore not dynamic. (Heated water affects room temperature, which affects the thermostat, which ultimately affects the heating of the water again.) He maintained that a system could be open only if it were non-mechanistic and were capable of handling the exchange of energy, matter, and information.

Open systems are indeed everywhere. Plant and animal cells are true open systems since they import energy in the form of chemicals, enzymes, or proteins and then utilize this energy to grow, multiply, and return waste products to the environment. Plants and animals, composed of trillions of smaller systems (i.e., their cells), perform these same functions in nature. Likewise, the entire ecosystem receives solar energy and distributes it across the planet, enabling photosynthesis to operate in plants and create oxygen from carbon dioxide. Next, the stored energy in plants creates the food chain, which is balanced by the existence of predators and prey. Solar energy also affects weather patterns through temperature and precipitation, both of which affect the food chain and plant life.

It may sound as if we have just described feedback loops, but what keeps the ecosystem dynamic and open is its ability to utilize (and transform) energy, matter, and information at virtually any place where interaction occurs within the system. For instance, predation can severely impact the populations of some species, which often opens a niche for new species of plants and animals. When a comet or asteroid impacted the earth 65,000,000 years ago and destroyed the dinosaurs, the environment was flexible enough despite global devastation to allow mammals to grow in number, size, complexity, and intelligence.

A high magnitude of order emerged from an equally high magnitude of chaos.

The extreme changes that can occur within a system are a demonstration of Chaos Theory. Chaotic conditions lie outside the normal limits of prediction or perceptual experience. (The odds of earth being hit by a comet are very slim, and yet scientists now know that such events happen at random intervals.) We will not go into the complex mathematical foundations upon which Chaos Theory rests, but what is important to understand at this point is that despite changes, a system can be restored to equilibrium. We will return to this idea shortly.

Social Systems

Because of man's ascendancy in the evolutionary process, his larger brain has produced the ability for advanced thinking, and with this cognitive functioning we see the invention of numerous social systems to meet man's needs. Families, businesses, manufacturing, governments, associations, hospitals, banks—all make possible the functioning of a highly complex society and all possess the dynamic attributes of open systems. A bank receives money from deposits and lends it so that the bank may receive more money again in the form of interest paid on loans. Loans are made for small business start-up, home construction, and automobile purchases, thus exchanging money with other systems. When people have housing, transportation, and jobs, they are once again in a position to deposit money in the bank.

We could trace similar patterns for *all* other cultural systems. In doing so, we would find that no social system is pre-eminent. No system could function, for example, without an educational system helping people to gain knowledge about their environment (and preparing them to function in other systems). Likewise, no large-scale social system could exist without transportation or communication systems to make possible the dynamic movement of energy, matter, and information. The healthcare system helps system components—you and me—to keep functioning so that other systems can be maintained. In other words, any system (or component of a given system) cannot be regarded as more important than another. This idea was central to the work of Ilya Prigogine.

Entropy and Ilya: Order and Disorder in Dynamic System

Prigogine was bothered by the contradiction between the Second Law of Thermodynamics, predicting entropy and disorder, and the evidence presented by evolution, which shows that the entire universe is characterized by *increasing* order and recognizable structure. Prigogine won a Nobel Prize for proving that dissipative structures,* especially life forms, harness energy in numerous ways to produce open systems that not only survive, but advance in complexity.

An important part of Prigogine's body of work addressed the idea of hierarchies. For example, elementary particles form atoms, atoms form molecules, molecules form macromolecules, and macromolecules form complex biological systems. Prigogine felt that it was a mistake to view one level of a system as more important than another. Instead, he argued that different levels are dependent on one another, with higher levels dependent on lower levels as much as lower levels are dependent on higher levels. This is completely in keeping with von Bertalanffy's General Systems Theory, which states that a system is a whole composed of interacting parts.

Biologically, all human cells compose the larger organism of the body, but the body's entire energy system can suffer considerable harm from a single virus or bacteria in any one of its trillions of cells. Additionally, when the immune system is functioning properly, white blood cells are manufactured to fight infection, thus alleviating fever and muscle aches throughout the entire body, enabling the organism to conduct its normal rou-

*Structures that drive away, dispel, or scatter, such as sunlight dispels darkness.

tines in the macro world (which is where the virus and bacteria are encountered in the first place). It is impossible to say that the body is more important than the cells or vice versa.

It is this non-linear, non-hierarchical nature of systems that makes them susceptible to abrupt fluctuations, which scientists call chaos. Dynamic interactions—even small ones—can cause unexpected consequences, but the same non-linear, non-hierarchical nature of a system makes possible the resumption of normal functioning by way of its holistic design. The stock market often defies the forecasts of even the most knowledgeable economists, but corrections and recoveries are also part of the market's overall structure since the market is never dependent on a single stock. Even a penny stock can outperform a blue-chip on any given day.

The very same principle, in fact, produces what the NFL calls "parity," expressed in an adage that is worth noting: "On any given day, one team can beat any other team in the league." Blue-chips and sports dynasties may dominate their culture, but as dynamic systems, they are subject to equilibrium.

A Matter Worthy of Reflection: Using NATI to Understand Systems

Just think about it: everything around us—all life and all social systems—evolved from simple one-celled organisms, which themselves evolved from strands of protein floating in the primordial oceans. According to the Second Law of Thermodynamics, there is no reason to believe that any complex structure should have developed, and yet here we are, discussing Natural Thinking & Intelligence. The movement toward growth and the actualization of potential is universal, and as I stated at the outset, this potential for development is written in the very structure of the cell as evidenced by Derald Langham's geometrical model. The thirteen intelligences that correspond to the thirteen stages of cell development also reflect this concept: no single part of any system is more important to its functioning than another. In other words, the principles of NATI form a *closed system* and are fixed, possessing boundaries and precise structure, with all intelligence assuming equal roles. They are non-changing and invariant and relate to seeing things exactly as they are. Note that besides the thirteen intelligences, we can also include as absolutes both Polarity and Virtue.

How we use our intelligences is a different matter, for they are non-linear and capable of interaction in an endless number of ways. The conditions of our inner matrices are variable and relate directly to perceptions as opposed to how things really are. In other words, NATI principles are objective, while our use of them is subjective. What this means is that Natural Thinking & Intelligence is composed of both open and closed systems. That is, Natural Intelligence is the closed system, while Natural Thinking is the open system. Our systems are always unbalanced, but by the same token, they are infinitely flexible, capable of restoring equilibrium to patterns thrown into chaos by the Great Restrictors.

Before proceeding, let us look at the Human Character Formula again (A + B = C) as a real example of a NATI system in which Awareness plus Belief equals the character of our Communication. What follows is a case study that illustrates generally how our NATI system works. In this study, we will see (a) how the absolute principles of NATI are affected by an individual's perception and (b) how the information, plus the energy/tension of Polarity, functions in a system capable of realizing potential: the human being.

A + B = Stanley: Example Of A NATI Cop System

Stanley was a client who consulted me regarding a product that one of his colleagues in Chicago had designed and developed, a new kind of device used by bioelectrolicysts. Stanley had originally developed a similar device himself and knew exactly how it was supposed to work.) He wanted to know whether or not this product would gain wide acceptance. I pointed out to him that this was an issue of creativity with an "outer characteristic" since the product required marketing acceptance. We therefore reviewed the Creative Group in trying to answer Stanley's question regarding whether or not the capsule would be accepted in the marketplace.

His Awareness was active and open—he was searching. He believed that the product itself had all the components of success and could do what it was supposed to do. Expression was also developed since Stanley had laid the groundwork for the product's introduction by contacting people he thought should be involved in the project. He couldn't move forward, however. His Belief was in question because, in his own words, "I just don't think it's gonna happen."

After further questioning, I learned that Stanley did not believe that *all* biopractioners would use his product. From there, we moved on to his Awareness and Belief about the Concept of the capsule. His awareness of the Concept was also high, and he once again expressed Belief that the product would work (i.e., that the Concept was sound), and yet he still didn't believe that it would gain acceptance.

Consulting the worksheets I had given him, we examined the issue of Polarity. He had indicated in his scoring (a one-to-ten number system is used) that his Belief relative to the Concept of the product was weak.

Accordingly, it is obvious from Stanley's remarks that he was extremely uncertain about the marketing phase of the product. Even more telling was his assessment that the product *might* gain acceptance, but *he* wasn't sure if he was the person to make it happen. The reason? Stanley's former partner had passed away two years earlier. As one of my associates in the session told Stanley, "It's just that you have not yet decided whether you want to work with a partner yet and then lose control." (Recall that the new capsule had been developed by a colleague in Chicago.) This mindset resulted in weak Belief. Because of the loss of his original partner, Stanley was feeling insecure and lost. *Internally, he did not believe in himself.*

What I want to emphasize for the moment is how important Belief was in Stanley's inner matrix. He had developed a very useful product that needed active, confident marketing. But while Stanley was not realizing his potential, it was not because of his product. The problem was within Stanley. *His Current Operating Procedure was static rather than dynamic.* Accordingly, Stanley's system could not interact with an even larger system (marketing) that would have resulted in the nationwide exchange of information about the capsule. A healthy Belief within his matrix had not been activated.

Stanley's case is a prime example of how NATI seeks to identify a weakness and use the tension between polarities—in this particular case study, active and inactive, passive and aggressive—to achieve a desired goal. As is the case with all open systems, equilibrium and balance can be achieved when we activate the necessary parts of our inner matrices to realize potential.

We have mentioned the importance of energy exchanges within open systems frequently in this chapter. In everyone's inner system, energy comes from the tension of Polarity. This is the key to change!

Summary

The ability to transform weakness into strength implies a dynamic process in which a person's intelligence and polarities can interact to achieve growth. This is a characteristic of systems theory.

An open system, as described by Ludwig von Bertalanffy, is comprised of components that can interact with each other and the environment after receiving an input of matter, energy, or information. A closed system is largely mechanistic and cannot interact with an exterior environment.

Open systems can restore equilibrium caused by random fluctuations, sometimes called chaos by scientists. In NATI terminology, a person's inner matrix—intelligences and polarities (all part of one's life philosophy)—are corrupted by rigid, conformist thinking. While the thirteen intelligences are absolute and fixed and therefore represent a closed system, we create open systems when we focus creatively on a problem or issue, bringing greater balance to our matrix. Natural intelligence, which represents our thirteen intelligences, is a closed system. Natural thinking, which represents how we implement the thirteen, is an open system.

Chapter Ten

Complex Adaptive Systems and The NATI Model

We will now analyze our system by comparing and integrating the thirteen intelligences with General Systems Theory.

A Brief Trip To Holland

Mathematician John Holland is a pioneer in the modeling and simulation of neural networks. In his work on Complex Adaptive Systems in biological organisms, he provides an important summary of what we have noted about systems and leads us to the next interface between NATI and GST. According to Holland, these are the most important points to remember as we proceed:[1]

- Each system is a network of many agents, or parts.
- Control within a Complex Adaptive System is highly dispersed. For example, there is no master neuron in the human brain. (This is yet another example that hierarchical control lacks validity.)
- Different levels of a system serve one another. The motor cortex of the human brain, for example, reinforces the visual cortex. The speech center serves various other centers within the brain, and so forth.
- Complex Adaptive Systems revise and rearrange their-components as they gain experience. The brain strengthens or weakens connections depending on encounters with the outside world.

- Complex systems usually provide niches that can be exploited by any of the system's components.
- Complex Adaptive Systems can anticipate future goals.

It is this last point that now deserves considerable attention, for the ability to predict future events is contingent on one crucial area of systems theory: symbols.

LUDCRETE: RIKU AND THE IMPORTANCE OF SYMBOLS

As we have seen, General Systems Theory was the brainchild of a biologist. This is not surprising since all scientists thrive on order. (One of the Core Human Dynamics of any good scientist is the drive or motivation to find order in his or her experiments, observations, and theories.) Experiments are carried out in methodical ways that yield a maximum of data that needs to be analyzed in order for scientists to reach conclusions. But exactly what are data? As we saw in RIKU, data are points of information and may be straightforward—the room temperature where an experiment took place—or complex, such as the amounts, sizes, or shapes of certain molecules. But these kinds of data are always represented symbolically in the mathematics of science. Indeed, without symbols, we have no data, no communication, and no science.

Since the communication of information is a basic principle of GST, symbols are crucial to the successful operation of any open system. Symbols represent a language, the transformation of concepts from the abstract to the concrete. The words you are reading at this very moment have symbolic meanings to which we must all agree if communication is to take place. Let's use the word "concrete" as an example. If I insist that "c-o-n" is pronounced "l-u-d," we get nowhere, for the symbol "ludcrete" will not activate any pattern recognition in our brain's speech and language centers.

When we consider that human language evolved in a rather arbitrary manner over tens of thousands of years, it is nothing short of miraculous that anyone learns to speak at all. A young child has no set of rules to follow when learning to speak, but somehow she ends up with the ability to express herself using over a hundred thousand words, which follow rules of grammar, syntax, and usage. The ability to organize complex rules of grammar and vocabulary is hardwired in our brains. Using trial and error, the child will eventually "get it right" because her brain is designed to understand symbols.

So how does General Systems Theory accommodate language or any other symbol? The answer is that symbols are representative and stand for something. They are freely created and transmitted through the learning process. *Symbols themselves form part of an open system, one that is living and organic.* This is why the little girl is ultimately able to use language. In her attempts to "get it right," energy and information are exchanged within the speech center of her brain.

It is hopefully clear at this point why RIKU is so important. Acquiring raw data correctly increases the chances that symbols will be understood and applied in a constructive way.

In ancient times, symbolism was more the province of mysticism—tarot cards and pentagrams—but the twentieth century has seen a new acceptance of symbolism because of quantum mechanics and theoretical physics. In talking of the Uncertainty Principle and whether or not an electron has location or energy, mathematical and linguistic symbols are absolutely indispensable when discussing matters that veer sharply into the realm of the abstract and theoretical. The same can be said for the Special and General Theories of Relativity. Without symbols, how can science even begin to discuss the interchangeable nature of matter and energy or the curvature of space, which is gravity? We begin to see more and more evidence that our physical reality is explained by a branch of science—quantum mechanics—that brings ancient metaphysical principles into play. We are left with a complete dependence on symbols to describe what we perceive to be real. In saying this, we have arrived back at the starting point, where we spoke of how the Tao regards creation as an expression of potential energy from an intelligent nothingness. If this is the case—and quantum mechanics lends strong credence to the notion—then the universe itself is merely one vast symbol for potential.

Power To The People

So this is our situation: We are symbolic creatures in a symbolic universe, and the most important thing we can do, in imitation of the universe itself, is to realize our potential. We can therefore say that the existence of symbols has some rather exciting consequences for us in the here and now, on the plane of everyday existence. First, given the non-hierarchical structure of nature, we can say dynamic interactions are now possible that give humans a mastery over space and time. This, in effect, enhances the meaning of the Core Human Dynamics of power. In the absence of linear notions of "before and after" or "cause and effect," humans can realize the same quantum leaps developmentally that subatomic particles experience physically. *When we bypass typical linear notions (following antecedents) and utilize absolute rules of nature (such as those of wholeness and integration), we come to a greater level of achievement.* In turn, this means that reasoning and intuition have the potential to completely circumvent the process of trial and error.

Humans are capable of moving past experimentation, though it remains an invaluable tool in both science and everyday life. (Figuratively speaking, we might say that our Intuitive intelligence has a little more breathing room.) Finally, and most importantly, symbols enable man to anticipate. It is this last feature that is an aspect of all open systems that makes it possible for individuals to address future actions. In Natural Thinking & Intelligence, we say that:

- A precise language of intelligences is in place so that...
- Our inner matrices may function in order to...
- Achieve clarity and understanding so we can...
- Set goals and enhance the future by...

- Solving problems and conflicts by...
- Addressing present weaknesses

This is the essence of NATI, as well as the reason we describe it as a system for decoding potential. NATI is a "systems format" for understanding the intelligence encoded in all of nature, including man.

A Vocabulary of Potential

NATI is, therefore, a real language, and you have already been introduced to the most important words in its vocabulary in the previous chapters. In learning the definitions of the thirteen intelligences, you were, for all intents and purposes, familiarizing yourself with part of the NATI dictionary. Below are the intelligences we have been discussing and their associated terms, called archetypes. These are synonyms or related interpretations for any given NATI principle. We sometimes use archetypes as alternate meanings in place of the intelligence term.

CREATIVE: *Origination, Plan*

- Focus: Awareness, Will, Motivation, Interest
- Belief: Concept, Content, Perception, Ideas, Image
- Expression: Development, Action, Personality, Realization, Character, Communication

ORGANIZATIONAL: *Formative, Structure*

- Models: Rules, Form, Laws, Identity, Support, Basics, Agenda, Principles
- Detail: Components, Individual, Segments, Separate, Unit, Sector, Segregate, Subdivision, Parts
- Priority: Measure, Levels, Degrees, Relatives, Emphasis, Intensity, Significance, Comparison, Judgment, Probability, Infinite, Dimensions
- Mirror: Reflection, Complimentary, Response, Interaction, Resonance, Feedback

- Whole: Total, Unity, Entirety, Overall, Collective, Group, Composite, Continuous, Generative, Connected, Integrate, Synthesis
- Order: System, Perimeters, Format, Pattern, Discipline, Process, Procedure

FUNCTIONAL: *Actualize, Accomplish, Doer, Dynamics, Achievement*

- Emotional: Feelings, Wants, Sensitivity, Desire, Excitation
- Physical: Material, Energy, Reality, Environment, Manifestation
- Mental: Intellect, Knowledge, Understanding, Choice, Thinking, Concentration
- Intuitive: Spirit, Sense, Essence, Innate, Psychic

It might be convenient to think of the above list as a type of architecture, for the term "architectonics" refers to the science of systematizing knowledge, which is exactly what Natural Thinking & Intelligence does. With faithfulness to the scientific method, NATI interprets and categorizes data so that meaningful conclusions may be drawn about our strengths and weaknesses. We do this always with the intention of altering our open systems to effect personal growth.

We also include the Human Character Formula as an important part of the NATI vocabulary since the "equals sign" (=) is the mathematical equivalent of a verb and, in a very real sense, connotes grammar. Saying that "our awareness, plus what we believe about our awareness, always equals what we are capable of becoming" is a grammatical, symbolic representation for decoding potential. (You will recall that I intentionally modeled this formula on the Pythagorean Theorem, which is part of the mathematical language of geometry). By means of NATI's symbolic systems language, we can anticipate (in the most positive sense) whatever future goals we choose. Symbols are valuable tools for making abstract goals concrete. Once symbols are correctly understood, our expression of those goals is made possible with the help of drive and motivation, creating a mindset directed toward development.

A Grammar for Success

Consider the following case study that succinctly illustrates how the basic language of NATI can point someone in the right direction. The chart below was made for a client who was enjoying a fulfilling career in a moderately sized company. To enhance his functioning, he was seeking a formula for successful leadership. We determined that his objective was indeed leadership. Noting that leadership belongs to the Creative Element of Expression, our mission was to come up with a formula for successful leadership that could be presented in an outline form. The intelligences, Polarity, and Core Human Dynamics produced the following mindset conducive to leadership.

Formula for Successful Leadership

- A leader expresses Awareness (Focus)
- A leader expresses Confidence (Beliefs)
- A leader expresses through Communication
- A leader demonstrates Organization (Structure)
- A leader demonstrates Order (Procedures)
- A leader expresses Judgment (Measurement)
- A leader seeks Details (Components)
- A leader demonstrates Teamwork (Unity, Totality, Inclusion)
- A leader demonstrates both sides of the picture—a balance—Polarity/Complementarity
- A leader expresses Energy (Vitality, Power)
- A leader expresses Desire, Sensitivity, Nerve (Emotion)
- A leader demonstrates Understanding, Intelligence, Common Sense (Mental)
- A leader expresses Spirit (Purpose, Vision, Drive)

The entire "grammar" in the above outline resulted in greater leadership potential for the client, who was able to access the abstract concept of leadership through the use of NATI symbols. This was accomplished simply due to the accuracy, validity, and comprehension of the multiple components of NATI and their definition of potential. In effect, science converts mystery into enlightenment.

The NATI Systems Language

The notion of universality goes back to the ancient Greeks and was a favorite topic of Plato. The philosophical pursuit of what is universal continued for hundreds of years until, in the twelfth and thirteenth centuries, the concept of universality centered on language. Great thinkers wanted a set of absolute symbols that everyone could relate to. Men like Liebnitz, Pascal, and Sir Isaac Newton tried but failed. Natural Thinking & Intelligence, however, represents that language! Using NATI principles, one is able to detect patterns that normally require in-depth analysis, introspection, time, or expertise. This is accomplished by an interpretive format, which is surprisingly easy to comprehend. (Derald Langham used to say that his concept was an *attempt to organize, synthesize, and synchronize data*).

Just as lawyers have their own language (i.e. laws vs. logic), so does NATI. The NATI language deals with factors such as direction vs. stagnation, objective vs. subjective, wholeness vs. separation, impersonal vs. personal, and following nature's path.

Here are some examples. Let's take the first set of factors, direction vs. stagnation. By direction we mean movement toward development, while stagnation describes things such as "the same old thing," victimization, or accepting the status quo.

With objectivity we interpret information as it is, without a particular frame of reference, as compared to subjectivity, which does possess a particular viewpoint. This is a major cause of distorted judgment and actions.

Wholeness here means integrating information and events into a comprehensive, all-encompassing concept of reality. This comparison of wholeness with separation refers to isolation, selectivity and self-orientation.

Being impersonal relates to detaching oneself from analyzing information or making a judgment. When it's your child vs. the neighbor's, impersonal says it doesn't make a difference—right is right—as compared to personal views where "my child is right no matter what." Understanding becomes more vital than emotion.

Lastly, our system defines itself by following the science of nature. An example of this science is the issue of complementarity. As a reminder, this deals with issues of opposition we all encounter and how we can mediate them. (For those who may be interested, I have included a list of twenty-two scientific theorems, which relate to NATI and are found in nature.)

This last matter of following nature relates to "letting go" versus "staying in absolute control." I realize that this is a most difficult state of mind to achieve. God knows I had trouble enough applying this principle of NATI systems language. But consider this: What convinced me and others that this is a valid and useful notion is that nature is a system that is greater than humankind, yet it includes it! The opposite is not the case. Issues in nature, such as its self-organization, are not interrupted by factors such as emotions, concepts, opinions, distractions, pain, pleasure, and the worst of them all, preconceived notions!

Nature, like potential, from where it comes, has no particular frame of reference.

NATI — A Journey of A Thousand Miles: Self-Actualization

Thus far, NATI has been described as a system in which a person focuses awareness on information in order to activate potential. It is a system that enables one to realize goals through higher levels of functioning. This concept of self-actualization was, in part, made popular in the 1960s and 1970s by psychologist Abraham Maslow, who attempted in his theories to create a scientific view of man. He did this primarily by articulating his now-famous hierarchy of needs, at the top of which was self-actualization. According to Maslow, self-actualization, or the development of our inherent potential, can be achieved only after certain "deficit needs" are met. These include the basic needs for nutrition, safety, love, and self-esteem. After these fundamental requirements are satisfied, we are in a position to achieve higher potential and develop skills or interests for our personal enrichment and the advancement of society in general. Maslow's theories have been instrumental in developing this scientific system of self-actualization. In the same way, Natural Thinking & Intelligence is a humanistic science, and as will be seen throughout this book, complements Maslow's ideas of self-actualization.

Maslow introduced eight factors that are necessary for self-actualization:

1. One needs to possess complete concentration.
2. One must realize that self-actualization is an on-going process.
3. One must possess an exemplary character and strong personality.
4. One must practice honesty and virtue especially when plagued by doubt.
5. One must make choices leading to growth (rather than being restricted by Fear, Ego, Self-deception, or Ignorance).
6. One must use his or her intelligence.
7. One must recognize peak experiences of joy and enlightenment.
8. One must have a strong sense of identity.

The first factor, concentration, naturally corresponds to the NATI principle of Awareness, and it is certainly no accident that Maslow listed this factor before all others. Awareness is akin to a valve that allows water to irrigate an entire field. Once irrigated, crops can grow in a normal, healthy manner. In the same way, once we are aware—once the doors of perception are opened, to paraphrase William Blake—we are free to grow by the application of the other principles. An individual who can focus his awareness will be someone with a strong personality and sense of self, someone who can choose virtue over selfishness, growth over fear. Indeed, fear may be regarded as the greatest impediment in any individual seeking higher achievements. Just as awareness opens the door for growth, fear totally closes off the possibility of realizing the peak experiences that enable us to savor life and its moments of transcendental bliss. It also blocks out any ability to function at the mystical, intuitive level, intuition being an extremely important kind of intelligence. When fear is eliminated, people are free to engage in frank, subjective discussions in a morally neutral and value-free environment. Greater amounts of information can be communicated because the threat of narrow-minded judgment has been eliminated by a higher level of functioning.

In short, fear prevents us from reaching most of the loftier goals we set for ourselves. That's why focus is so important, for without awareness and focus, we cannot identify the fear-based defense mechanisms that limit our potential for growth, natural intelligence, and enlightenment. Only by recognizing our defenses can we possibly find the courage to give them up. Paul Dirac, a pioneer in the field of quantum physics, once proclaimed that great breakthroughs always involve the sacrifice of some prejudice. This is true not only in science, but also in the developmental stages of every human being.

Omega

Management guru Peter Drucker has said that, "Unless information is organized, it is still data,"[2] which corroborates what we have already noted in relation to RIKU. Drucker further states that once information has been assembled in the workplace, meaningful communication takes place when workers have a focus on something, with the best focus being a common task or challenge. Drucker goes on to say that

> "No two executives . . . organize the same information the same way. The information has to be organized the way the individual executive works, but there are some basic methodologies to organizing information...This key event [of organizing information] may have to do with people and their development."[3]

What NATI proposes is that Drucker's corporate model for productivity and development holds true for individuals, groups, sub-cultures, nations, and indeed, the entire world. This is the ultimate mission for Natural Thinking & Intelligence: to achieve the highest degree of development possible. In Maslow's terminology, the goal is for people to become self-actualized.

At this point, one might be inclined to say that realizing potential is all well and good, but it is simply an ideal, an outworn cliché reserved for teachers and pop psychology books. I can assure you that NATI transcends traditional ideas of self-improvement, for Natural Thinking & Intelligence looks far beyond the cultivation of a single virtue by just one individual at an isolated moment in time. Unlike all how-to books in the psychology section at the bookstore, this book presents a comprehensive system that uses every type of human intelligence in a holistic fashion so that everyone will be free to develop in any number of ways and in all possible situations. Even more importantly, NATI seeks to help the entire species move toward a destiny that some might consider to the province of science fiction.

Teilhard de Chardin believed that humanity was heading toward a final point in the evolutionary process, a destination he called the Omega Point. He believed that all of mankind was destined to be assimilated into a global consciousness, an endpoint that can be attained only after individuals strive to perfect themselves. In The Divine Milieu, Teilhard de Chardin expressed the belief that each and every act performed by an individual, no matter how small or seemingly insignificant, is capable of merging human experience with divine consciousness.[4]

By analogy, we can say that by intelligently focusing on every bit of information acquired from raw data, no matter how small or seemingly insignificant, we can strive toward the ultimate goal of perfection. While this may seem to be a daunting task, one needs to remember the Buddhist aphorism that "It is the journey, not the destination" that is important. Still, it is not until we have an inner image of the perfected human psyche that we can begin that awesome journey to ultimate fulfillment. It is paramount in our journey that we have perfection as our clear goal, even if it exists only as a mental construct. We may never see the Omega Point in our own lifetime. Whether such a cosmic event will be attained in a thousand years or ten thousand years is unknown. But if we take every opportunity to improve ourselves, humankind itself will be poised for an evolutionary leap that transcends the mere biological or physical development of the species usually implied by the term "evolution."

The Chinese proverb says that, "A journey of a thousand miles begins with a single step." With NATI, every time we form data into information and apply it to a situation with awareness and wisdom, we take a step toward a realm of unimagined possibilities.

The Absoluteness of The System; Birth of the Science of Achievement!

It's a pretty grand statement to say that something is absolute. We are talking about perfection in a very real sense. So we need to approach declaration of absoluteness with critical rigor. Let's review this contention with each of the 13.

The Whole	The greater the whole the better. The more details, systems, models, feedback and measure you can employ the better you become.
The Mirror	The dislikes you reflect are your inner faults, weaknesses.
Feedback	We use information to improve things.
Measurement	Everything is relative. Relativity is a valid theory.
Details	Everything is a part of a whole; even the wholes. Also separation is a part of something else.
The Process	Nature is self organizing; it has two systems we all utilize—open and closed.
The Rules	Fundamentals are essential and totally reliable. They are dominant.
Polarity	The world is composed of positive/negative.
Awareness	We all have this quality.
Concepts	We all have beliefs.
Expression	We communicate whatever we focus on and our belief about that equates to how we express it.
Potential	Contains all possibilities.
Development	The one thing which will ensure an ultimate positive direction individually and collectively.
The four functional intelligences—Physical, Mental, Emotional, Intuition	These are the only ways by which we can function.

An Interactive, Universal System

The validity of a system can be tested by its consistency, flexibility, and universality in its application. The following scenario demonstrates how the three general classifications of NATI accomplish this end. It does so by establishing valid axioms concerning the mechanics of NATI.

- **Axiom # 1 — Function x Organization = Planning**
  To proceed to Planning we need to incorporate functioning and organization principles.

- **Axiom # 2 — Planning x Functional = Organization**
  To Proceed to Organization, we need to incorporate Planning and Functioning Principles.

- **Axiom # 3 — Organization x Planning = Functioning**
  To proceed to Functioning we need to incorporate Planning and Organizing Principles. Any and all combinations of the three will always bring the same results. Therefore, they are dependable and viable.

The Flexible NATI Structure — An Aristotalean Viewpoint.

Aristotle often spoke of "the Unmoved Mover", a concept of the "first cause" for moving the universe along, so to speak. What he meant by this was, basically, that there were three phases of universal structure; the fixed component, the constant component and the realized (resulting) component. Here we see the A+B=C theorem once again with A = fixed, B = constant and C = realization. But there is more to this notion of unmoved mover and first cause. Here's a graphic of his thinking.

Fixed Element	**+**	**Constant Element**	**=**	**Resulting Element**
1		1		11
1		2		12
1		3		13
1		4		14
1		5		15

NATI works in the same fashion. Here is one graphic to demonstrate.

Fixed	**+**	**Constant**	**=**	**Result**
Awareness		Awareness		Awareness of Awareness
Awareness		Belief		Awareness of Beliefs
Awareness		Expression		Awareness of Expression
Awareness		Model		Awareness of Models
Awareness		Process		Awareness of Process

This procedure can be carried on ad infinitum. The NATI Principles can also be interchanged within the three structures. Additionally, any one of them situated in the fixed component becomes the "first cause". That is to say they are the very factor which everything that follows is based upon.

Summary

There are several factors that enable us to follow nature. The simplest for the average person is the thirteen principles of NATI. The most difficult factor is recognizing nature's principles in science. Other factors are development as a life purpose as well as seeking out our weaknesses and dealing with them. There is one other factor to consider, the pursuit of virtue.

The Complex Adaptive Systems we examined all share the common characteristic of anticipating future goals through symbolism. Identifying goals is paramount in mind science, the primary goal of which is development. The most common form of symbolism is language, which enables information to be communicated in healthy, open systems as described by von Bertalanffy. NATI itself is a universal language, its vocabulary comprised of thirteen intelligences, each with archetypal qualities, and polarities.

The NATI Systems Language breaks a simple idea into NATI terms and then translates it for the purpose of understanding issues in clear, rapid fashion.

As we will see in the next chapter, any viable scientific theorem can be utilized as a NATI systems language device.

The focus on and organization of acquired information activates the potential energy that is dormant in every human being. Focus is one of the many ways that NATI helps people achieve a higher level of functioning, which is called self-actualization by psychologist Abraham Maslow.

When enough people begin to focus on the development of potential in their lives, humanity as a whole will be advancing toward what some philosophers call the Omega Point, a place where humanity will have evolved a global consciousness. It is NATI's goal to help both individuals and society reach new heights. At the same time, it is the Science of Achievement which is the vehicle!

Nati Model of Existence

A Diagram of the Nothingness to Actualization

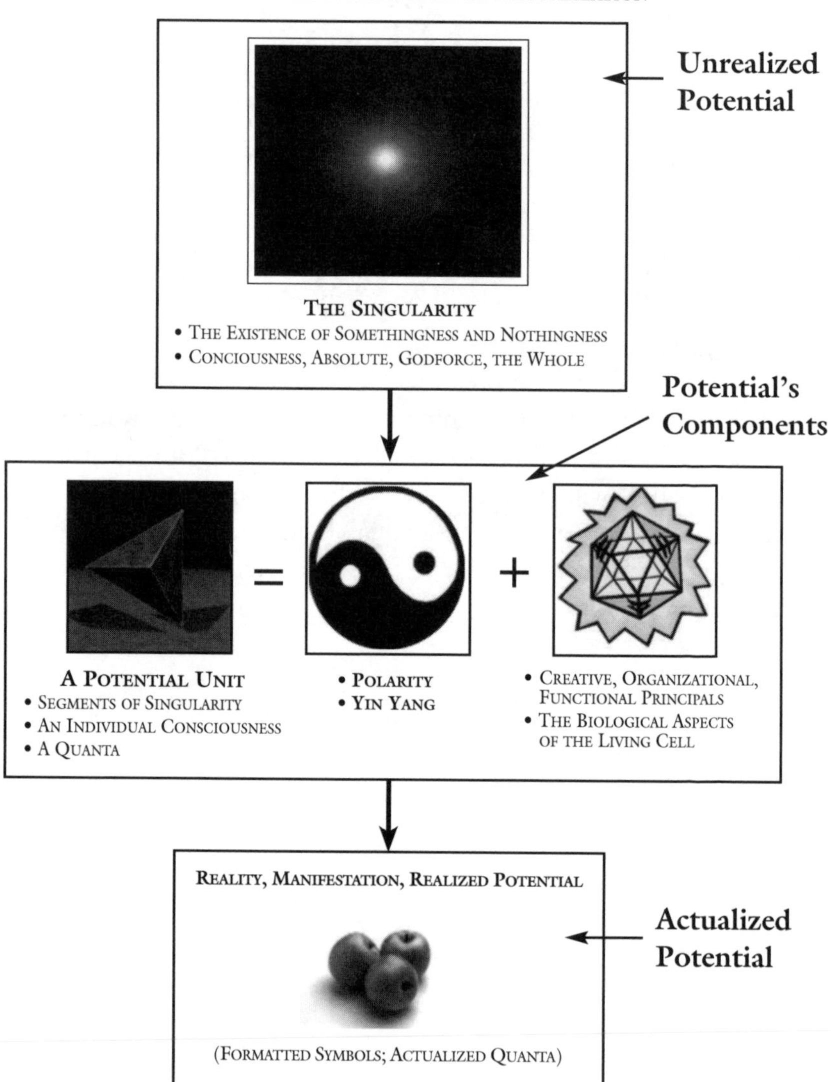

Chapter Eleven

Physics and Philosophy of Whole Systems

It is important that we formulate a picture of how Systems Thinking works. From our examination of systems, we know that our thirteen intelligences are fixed. Each intelligence and its Polarity is an absolute, and yet taken together they form a dynamic pattern when they interact. But as we have also seen, a system is not so much a "thing" as it is a whole or a pattern of organization. As Lawrence S. Bale says:

> Systems are comprised of a unified pattern of *events*, and their existence, as well as their character, is derived more from the nature of their organization than from the nature of their components. As such, a system consists of a dynamic flow of interactions that cannot be quantified, weighed or measured. The pattern of the whole is 'non-summative' and irreducible.[1]

If you look back at our theory of potential, you may see how this statement matches up. Potential is truly non-summative and irreducible. By this we mean it is too expansive, too all-inclusive to be summarized. It is also absolute by virtue of the fact that there is nothing beyond it.

Some physicists, such as David Bohm, have gone so far as to say that not only are system components interdependent, but they are ultimately *all the same thing* and that the separateness of all objects is an illusion. What is meant by illusion is not that it is unreal; rather, it implies that there is another and greater underlying reality. Bohm, for example, believed that everything in existence is part of a higher, abstract dimension, a view shared by Prigogine. This notion, in essence, is the offspring of General Systems Theory and is known as Holographic Theory. Since the views of scientists once again bring us to a point where science meets mysticism, we need to explore Holographic Theory before examining how the human brain itself is perhaps the ultimate hologram and is responsible for the most important task of all: decoding potential.

A Horse is a Horse is a Horse: What a Hologram is

First, we need to explain what a hologram is. We see holographic pictures in tourist shops, on baseball cards, and in futuristic science fiction movies. The technology required to make a hologram, however, is relatively simple. A hologram is produced when a laser beam is split into two distinct beams, one bouncing off an object and the second serving as a reference beam. In other words, the second beam can be examined "in reference" to what is happening to the first. When the second beam is aimed at a photographic plate, a curious thing happens. A wave interference pattern is formed by the two beams that bears little resemblance to the object struck by beam number one, and yet the resulting image contains all the information required to recreate a picture of the object. Even more astounding, every portion of the photographic plate, no matter how minute, contains an image of the entire picture. If you used this holographic process to take a picture of a horse and then cut out a tiny portion—an ear or a hoof—you would still be able to detect an image of the entire horse on the smaller section, although its resolution would be fuzzy depending on the size of the section. In other words, each piece of the cutout is encoded with information from the original to produce an image of the whole.

This process, which can be replicated in almost any photographic laboratory, has profound implications for science and philosophy. We have already noted that each cell in the body contains complete instructions for forming any other part of the body (which has now become front page news because of cloning), but the implications are equally profound for mind sciences such as Natural Thinking & Intelligence. As we shall shortly see, holographic principles are responsible for how the brain decodes potential—that is, creates yu, the manifestation of the absolute, or the pregnant void. In fact, we see in Holographic Theory a supreme act of synthesis since it explains why we are able to perceive anything at all.

Just as we examined von Bertalanffy's ideas to understand systems better, we will look at the work of pioneers in Holographic Theory, such as David Bohm and Karl Pribram, since we are now at the threshold of understanding how the complete NATI vocabulary translates potential into reality.

Drops of Glycerin, Pearls of Wisdom: Bohm's Holographic Theory

In *A Holographic View of Reality*, David S. Walonick states that, "According to Bohm's theory, the separateness of things is but an illusion, and all things are actually part of the same unbroken continuum. Holographic theory is an extension of General System Theory because it recognizes that the boundaries of a system are an artificial construct."[2]

This idea that boundaries are illusory gives new meaning to the non-hierarchical nature of systems. The dynamic interaction among non-linear system components is actually made possible because the components are part of a continuum for which we must actually redefine "interconnectedness." We normally think of connections as "joints" where physical matter from different objects meets, such as lengths of pipe in a plumbing system. In geometry, connections are represented by vertices of different line segments. In Bohm's theory, however, a "connection" is entirely different. The Cartesian view of matter intersecting along X, Y, or Z axes becomes a mere convenience for describing the macro world. For Bohm, the true nature of connections is more accurately described by a concept called "enfoldment."

One of Bohm's early (and quite famous) experiments illustrates this idea in a relatively simple fashion. Bohm placed a small glass cylinder inside of a larger, concentric glass cylinder, filling the space between the two cylinders with glycerin. (Glycerin is a clear, highly viscous liquid.) Next, a single drop of dark ink was placed in the glycerin while the inner cylinder was turned with a crank-handle. The drop of ink was stretched into a very thin thread that eventually disappeared altogether. The obvious conclusion was that the molecules of ink were thoroughly mixed into the glycerin in random fashion and therefore irretrievable by any process. This was not the case, though! When the handle was turned in the opposite direction, the thin, dark line reappeared. With even more turns, the line reassembled itself into the original ink drop. (An interesting aside is that Bohm got the idea of using the cylinders after seeing a similar device on TV, a classic case of modeling.)

From this experiment, Bohm realized that the drop had been *enfolded* into the glycerin when it disappeared. It was *unfolded* when it appeared again after the cylinder was turned in retrograde fashion. For Bohm, this model explained the nature of physical reality. He believed that electrons are non-local, by which he meant that they are normally enfolded into a higher dimension and not confined to a specific area.[3] They appear as physical matter when they are unfolded in what we perceive to be normal space-time. This happens continuously, so that the unfoldings occur adjacent to one another, producing continuous movement in what we know as the three-dimensional world. In Bohm's terminology, this higher dimension into which everything is enfolded is called the implicate order.[4] Note that in NATI terms, "implicate" relates directly to integration and synthesis. The explicate order refers to physical reality that unfolds from the implicate. Rather than being solid matter, the world we see around us is more of a projection from the abstract realm into which everything is enfolded and unified.

Just as there is dynamic movement and interplay between elements in a system, Bohm believed that there was a constant interplay between the implicate and explicate orders, which he termed holomovement.[5] He thought that existence was an undivided whole that was in a constant state of flux. In this model, mind and matter are not separate, nor is living and non-living matter. Indeed, even matter and empty space are not viewed as separate in Bohm's holographic theories. Independent research indicates that there is something called zero-point energy, a quantity of energy contained in a single centimeter of space that is greater than all the energy in the observable universe.[6]

In other words, all energy is everywhere at the same time.

The Horse Revisited: The Holographic Nature of Reality

To say that everything in the universe is non-local and enfolded into a higher dimension is exactly what is implied by the complete image of any holographic picture being present in any *part* of the picture, as was the case with the horse. The image of the horse is unfolded into all parts of the overall hologram. What Bohm is saying is that this kind of model exists on a much grander scale, with every part of the universe containing every other part. In "Lifework of David Bohm," Will Keepin gives us a remarkable example of how to visualize this.

> Imagine yourself gazing upward at the night sky on a clear night, and consider what is actually taking place. You are able to discern structures and perceive events that span vast stretches of space and time, all of which are, in some sense, contained in the movements of the light in the tiny space encompassed by your eyeball. The photons entering your pupil come from stars that are millions of light years apart, and some of these photons embarked on their journey billions of years ago to reach their final destination, your retina. In some sense, then, your eyeball contains the entire cosmos, including the

> enormous expanse of space and immensity of time—although, of course, the details are not highly refined. Optical and radio telescopes have much larger apertures, or "holographic plates," and consequently they are able to glean much greater detail and precision than the unaided eye. But the principle is clear, and it is extraordinary to contemplate.[7]

Bohm went so far as to say that there were *several* layers of implicate order, which he termed superimplicate. This is related to a field of physics called quantum potential, which we will not go into since its complexity is beyond the scope of our present discussion (although we *will* see a quantum Potential Intelligence Matrix in action). What is important is that these superimplicate orders, in Bohm's view, are parts of feedback loops, with higher dimensions contributing to the lower and, conversely, lower dimensions contributing to the higher. This, of course, is exactly the case with non-linear open systems. By extrapolation, we can therefore say that all such systems display holomovement.

Another example of the holographic nature of reality is the thirteen principles/intelligences of NATI. Virtually every important theoretical concept we have presented thus far is synthesized in this concept. We could say that all other information is unfolded from this concept, or we could just as easily say that all information from other concepts could be enfolded into this one.

We can also reiterate David Bohm's idea at this point that the similarity of differences is related to the difference of similarities. The various chapters you have read have approached certain topics from different angles (differences), but they have all been describing the same thing after all is said and done: the idea that our thinking and intelligence are reflections of nature itself (similarities). Furthermore, nature is reflected in the geometry of cells just as it is reflected in the human organism. There is nowhere we can go where NATI principles do *not* exist. The intelligences are enfolded, however, into the implicate order, where the Parts (differences) merge back into a Whole. In effect, the parts, while separate, become similar by virtue of becoming members of a whole. This reflects the organizational intelligence of Wholeness, or Synthesis. It is through the intelligence of Synthesis that we will ultimately see how the Potential Decoder Matrix functions in all situations.

OM AND BOHM: HOLOGRAPHY AND TAOISM

The implicate order can also be understood from the perspective of Taoism and the concept of the living void, also referred to scientifically and philosophically in these pages as potential. The implicate order is therefore the same as *wu*, while the explicate order—the manifestation—is the same as *yu*. Indeed, we can now say that the terms "manifestation" and "holomovement" are synonymous, for they both express the dynamic flux that is responsible for the continuous unfolding that produces reality. In effect, from the void (potential) comes the actualization of reality. Creation or creativity is, then, an ongoing process. This also happens to be why, in the course of my search in earlier years, I found it easier to deal with issues by first thinking on an abstract level before using symbolic language to express ideas in more concrete terms. To access the abstract—the implicate order—is the true meaning of transcendence. Whether we speak in terms of systems or quantum mechanics, to say that cause and effect are arbitrary is to move past the constraints of space-time and operate on a level where existence is a continuum rather than a sequence of events. This is precisely why oriental philosophy and mysticism stresses the ideas of oneness and unity. Ultimately, we are going against nature when we attempt to parse reality into various segments to which we attribute too much meaning. Parts certainly exist, and I am not saying that they are unimportant. Parts is, after all, one of the thirteen intelligences in the universe. But to decode potential properly, we must be able to use holistic or abstract thinking, which is composed of Parts. If the development of potential is our final goal, we must remember that potential exists as a unified field of energy in the implicate realm, not as any "thing" that is separate.

We shouldn't be surprised, therefore, to learn that David Bohm himself went beyond the bounds of traditional science to gain insights on matters such as existence, thought, reality, and truth. He met with several spiritual leaders, including the Dalai Lama and Krishnamurti.[8] By equating the superimplicate order with the Taoist void, Bohm was criticized sharply by many colleagues for moving from the empirical to the mystical. Bohm, however, believed that moving beyond the empirical was entirely warranted. Indeed, he was very moved by suffering and pain in the world, and in *Science, Order, and Creativity* (co-authored with F. David Peat), Bohm stated that

> What is needed today is a new surge that is similar to the energy generated during the Renaissance but even deeper and more extensive; . . . the essential need is for a loosening of a rigidly held intellectual content in the tacit infrastructure of consciousness, along with a melting of the hardness of the heart on the side of feeling. The melting on the emotional side could perhaps be called the beginning of genuine love, while the loosening of thought is the beginning of awakening of creative intelligence. The two necessarily go together.[9]

Although a man of science, Bohm advocated a more emotional approach to mind science (he was clearly using his Emotional intelligence here) in order to find a middle way that could awaken creative intelligence, which is a core objective of Natural Thinking & Intelligence. Notice that he is not advocating one approach over another, but rather is urging a "loosening" of conventional thought. He clearly wished to use the tension of polarities and the Principle of Complementarity to find common ground between Emotional intelligence and Mental functioning, which are so often at odds in science. In Bohm's thinking, the two simply "go together".

Neological Dialogue: Metaphysics And Science

Bohm was not the only scientist attempting to "loosen" scientific thought. Throughout the 1970s, physicists, biologists, and neurosurgeons were taking a careful look at metaphysics in an effort to explain the hard data of science. The experimental facts of science, from physics to physiology, only seemed to make sense if there was some unifying or transcendental ground underlying the explicit data.[10] These researchers believed that this timeless, boundless realm—the "Godhead"—had been universally described by the world's great mystics and sages, whether Hindu, Buddhist, Christian, or Taoist. I have personally found this to be the basis of the Uncertainty Principle; that is, there is an Absolute that cannot be measured or conceived. This was the central issue that prompted a Neological dialogue, and the interested reader can find a wealth of information on this subject in the classic by Fritjof Capra, *The Tao of Physics.*

Science now provides a non-doctrinal way to approach the idea of transcendence that was once the sole territory of organized religion.

Parallels in Psychology: Jung, Maslow, and Bohm

Jung's idea of the Collective Unconscious tallies quite well with Holographic Theory. Jung believed that certain dreams and symbols were universal and therefore were present to some degree in all times, places, and cultures. Using Jung's theory, we can say with validity that dreams and synchronicities might represent brief moments when humans are able to glimpse (or have access to) the implicate order, which is normally beyond the brain's perceptual threshold. Many researchers believe that by denying higher orders of reality, we have unwisely severed ourselves from the meaning inherent in deeper orders of existence. Perhaps if we allowed ourselves to be more in touch with the abstract, we could better synthesize the random events of our lives into a more meaningful holistic experience directed toward achievement.

Regarding this last point, this is exactly what Abraham Maslow advocated. By using virtue and intelligence to overcome fear and ignorance, we can enjoy peak experiences, moments that border on transcendence as man becomes more and more self-actualized. A willingness to try new things in Bohm's spirit of reaching beyond the constraints of traditional thinking has the potential to minimize Fear, Ego, Ignorance, and Self-deception. *Whatever terms we use—the living void, a higher dimension, the implicate order, the Absolute—we are always speaking of potential. The drive toward development is universal.*

Parallels in Science: Einstein, Bohm, and Bohr

While science is not always sympathetic to the philosophical implications of research, certain aspects of quantum mechanics nevertheless bolster Holographic Theory. The foremost example is the EPR Paradox (the letters stand for the last names of Albert Einstein, Boris Podolsky, and Nathan Rosen). The paradox involves the measure of two particles traveling away from each other. The polarization of both particles is always the same regardless of the distance between them when any observation is made.[11] Change the polarization of one, and the polarization of the other changes simultaneously. This was precisely what made Einstein so skeptical about quantum theory, because the change in polarization was instantaneous and implied that "something" traveling faster than the speed of light was connecting the two particles. This would violate, of course, the Special Theory of Relativity. It was Neils Bohr who argued that Einstein was wrong to view the particles as separate to begin with. As we have already seen, Bohr viewed the particles as complementary, as parts of a unified whole.[12]

The EPR Paradox was originally a thought experiment performed by Einstein and his colleagues. Subsequent experimentation by Alain Aspect in 1982 has proven the paradox to be true.

(Aspect heated calcium and then allowed resulting photon pairs to travel in different directions.[13] Polarizing one photon caused an immediate polarization of the other.) We have spoken previously of non-local intelligence, but the EPR paradox experiments show conclusively that there is such a phenomenon as non-local matter as well. This is in complete accord with Bohm's postulate of an implicate order. Holomovement accounts for the paradox through the constant enfolding and unfolding of the electron.

Just as the picture of a horse is present everywhere in its holographic representation, matter and consciousness are present everywhere in a holographic universe. Order is everywhere—we simply cannot perceive it. As Bohm himself said, the very idea of chaos may be an illusion. Einstein, too, certainly believed in the innate order of the universe, regardless of his misgivings over certain aspects of quantum mechanics. Einstein and Bohm, in fact, held many conversations on what might lie beyond quantum mechanics, although Einstein certainly did not become involved in mystical inquiry as did Bohm. He did believe, however, that a cosmic code of order was written into the universe.

Whether we hold to Einstein's cosmic code or Bohm's implicate order, we are left with a reality that can be understood and deciphered, and this is why the organizational principles of Natural Intelligence & Thinking are so relevant. They are capable of organizing any data at any time. The intelligences are universal, enfolded into the very fabric of existence.

Parallels Everywhere: NATI's Whole Picture View

NATI offers us a "whole picture" of the world that recognizes and categorizes patterns and brings order to life. We can expand our parallels even further through the following chart. If all knowledge is enfolded, then the following similarities can be explained in the sense that all information described in the systems below exists as a whole in the realm of potential. In the explicate world, the similarities (once again) have differences, but

Chart of Philosophic & Scientific Components of NATI

SYSTEMS	THREENESS	FOURNESS	POTENTIAL	POLARITY
Philosophy	Motivation Belief Expression	Earth Air Fire Water	Unmoved mover	Duality Mirroring
	Creative Qualities of Pythagoras	*Functional Qualities of Greeks*	Concept of Aristotle	*Ancient Chinese concept of Knowing*
Psychology	Awareness Concepts Personality	Physical Mental Emotional Intuitional	The nature of human potential	Law of Complimentarity
		CarlJung	*Maslow, Feldenkreis*	
Science	Pulse Wave Spiral	Weak Strong Electromagnetic Gravity	Potential Energy	Opposites
	Three parts of cell division	*Four forces*		
Nature	Length Width Breadth	Earth Air Fire Water	Balance	Inversion
	3 Dimensions	*4 Elements*		
Math	X Y Z *Cartesian coordinates*	Plus Minus Multiply Divide	1	1/N

the differences could easily be enfolded again to a point where they existed in implicate form as similarities.

All of the correlated concepts in the cross-disciplines listed in the table are, at some point, expressing the same ideas. Math, for example, reflects nature. To say that there are three dimensions in nature is no different than saying there are X, Y, and Z axes on an algebraic graph. And as we have already seen, Pythagoras himself believed various formulas and theorems reflected more basic, philosophical truths. Hence, we can model the Human Character Formula of $A + B = C$ on geometry, knowing that the same precision, symmetry, and truth that applies to the Creative Group of intelligences also applies to a right triangle.

There is unity and order in the holographic universe, and where there is order, there is the potential for not only understanding it, but for using its principles. As you can see, this cross-disciplining characteristic is central to our Quantum Potential Matrix for decoding potential.

Summary

Reinforcing von Bertalanffy's idea that a system is not just a "thing" so much as a pattern of interaction, David Bohm used Holographic Theory to view system components as not only nonlinear and non-hierarchical, but as part of what he called the implicate order. In this view, system components are dynamic because, at a deeper level of reality, they are unified.

The NATI matrix is essentially holographic in nature. Using holographic principles, the human brain decodes the absolute and manifests potential in what Bohm called the explicate order, which is ordinary space-time.

Holographic Theory also reinforces the Science of Achievement concepts since all thirteen intelligences are always working simultaneously and are manifestations of natural laws unfolded from higher dimensions of reality.

PART THREE

THE PRINCIPLE OF COMPLEMENTARITY

Chapter Twelve

The Thirteen Intelligences and Whole Brain Thinking

We will now move from Whole Systems to Whole Brain Thinking and the connection to the thirteen intelligences.

In one of the more startling examples of synchronicity in science, Karl Pribram was working on his holographic theory of the brain while David Bohm investigated the holographic nature of the entire universe. Bohm's work, of course, subsumes Pribram's inasmuch as the human brain is a part of the universe. Pribram's work deserves special attention, though, since the brain is our ultimate decoder of the absolute in the more literal, physiological sense.

The story actually begins with two of Pribram's mentors. Pribram was a student of Wilder Penfield, who claimed in the 1920s that memories were stored in specific locations within the brain. He came to this conclusion while performing brain surgery on epileptic patients. When areas within the temporal lobes were stimulated, Penfield noticed that his patients relived certain past experiences in a vivid manner, as if they were actually living the event for the first time. Penfield said that memories were therefore located on memory engrams within the brain.

Karl Lashley tried to verify Penfield's findings in the 1950s by training rats to run through a maze. Contradictory to Penfield's original experiments, Lashley found that the rats' ability to negotiate the maze remained intact even when portions of their brains were removed. In fact, the rats were able to run the maze regardless of how much brain matter had been excised.

Pribram was intrigued by Lashley's research and found that humans also did not suffer memory loss when patients (trauma patients with relatively severe injuries) lost a part of their brain tissue. Their memories were sometimes not as sharp, but they were nevertheless intact. This led Pribram to conclude that memory was not localized after all, but rather was spread throughout the brain.

This is remarkably like the nature of a hologram. Just as a picture of the whole is retained in any part of the hologram, no matter how small, memories appear to remain intact in smaller and smaller sections of the brain. The memories grow fuzzier, just as smaller holographic images grow less distinct, but they are still present after a reduction of brain tissue. Pribram noticed the physiological similarity to holographic theories immediately upon reading about holograms in *Scientific American*.

Several models have since been put forward to describe how the brain operates holographically. We saw in our discussion of Intuitive intelligence that physicist Fred Alan Wolf believes that waves pass through glial cells, forming oscillations corresponding (in the research of Renato Nobili) to Schrödinger wave energy. Pribram held a similar view, saying that waves cause interference patterns as they move across neurons, axions, and dendrites within the brain (a seemingly scientific basis for visualization and prayer).

David Walonick gives a more figurative description of brain holography. He states that, "In the brain, past experience might serve as the reference beam. New incoming information is combined with the experiences (memories) of the past to create an interference pattern." With new information constantly arriving and changing the interference patterns,[1] we perceive normal everyday events unfolding in a recognizable, linear fashion.

For his part, Bohm felt that the brain operated as a kind of frequency analyzer that could decode impulses from the implicate order. This is similar to Pribram's view. Pribram hypothesized that the solid world we perceive is an illusion, a mere convenience allowing us to carry on certain actions. In short, the brain may be the lens through which we view the world so that we can make some kind of sense out of it. Pribram even suggested that if we saw reality without the benefit of the mathematical computations performed by our brains, we would perceive a world of pure frequency, without time or space, just events. Might this computational process be the distinction between the material world and the quantum world?

If this is the case, one might ask why everyone perceives the same time-and-space reality. The answer is that we are collectively influenced to register certain common mathematical functions in our brains.* If people are intoxicated or psychotic, however, their reality is significantly altered.

The Heart of The Matter: Whole Brain Intelligences

Getting to the heart of issues through whole brain processing is what Natural Thinking & Intelligence is all about. There are some fundamental differences between logical and whole brain thinking we should look at.

Logic says, "if this, then that." It is sequential and ordered. It basically makes sense of things as it goes along. If you don't know a fact, you quit, make an assumption, or postpone judgment. The proliferation of the latter by so many of us so much of the time demonstrates a weakness in this process. Further, if facts are not available and assumptions are utilized in their place, chaos and disaster can emerge. This is also a typical outcome in deductive thinking. One of the greatest restrictions in logical thinking is that it is based on known information. If it's not in a book, or commonly known or accepted, logic probably will not get you to the answer. Leaps of faith are not accepted.

Contrast logic with whole brain thinking, specifically in NATI terms. First, let me say that in my experiences over the last twenty-five years, whole brain thinking is more natural and is closer to the way the brain actually works. A key difference

*In certain circles, this is called the "collective consciousness."

between the two is that one utilizes given data and builds a sequential series (logic), while the other utilizes a structured system to examine data. This means that one can take any number of steps to find answers. In effect, results are based upon systems structure rather than qualified information. For example, if I were learning how to hit an overhand tennis shot step by step, I would look like a mechanical robot at first. But once I got into applying the information and really swung the racquet, I would be realizing a system structure.

Does this mean that logic is necessary for whole brain thinking? Not really. The fact is that no one really has all the physiological, neurological, and cognitive facts necessary to arrive at a logical conclusion about the action of hitting a tennis or golf shot. In this case, we take what we get, then act.

Einstein is a good example of a whole brain thinker. His famous leap of intuition brought him to $E=mc^2$.

The Science of Achievement accomplishes whole brain thinking by approaching a focus from twelve separate but connected parts. It then utilizes the one (or ones) that connects with the focus and continues on from there. For instance, if we focus on a problem and can't solve it logically, in NATI terms we would look at the problem from the following perspectives:

- Beliefs about it
- Its character
- Rules concerning it
- Its procedures
- Its priority
- What it is reflecting
- Its details
- The whole picture involved
- Mental factors
- Physical factors
- Emotional factors
- Intuitive factors

This is a far cry from being stuck in a one-dimensional process, dependent-upon questionable and/or limited data. With whole brain thinking you can start at any point and proceed. That is not the case with logic.

It is interesting to note that whole brain thinking actually incorporates logical and deductive thinking, while the converse is not at all true.

One of the beauties of whole brain thinking is its capacity to deal with opposites. Literal and figurative, positive and negative, personal and impersonal are all appropriate within whole brain thought. This enables complementarity as well as "big picture views" of matters.

The function of a lens is to focus. Through the Science of Achievement, we can focus on our intelligences, weaknesses, and polarities for the purpose of development. In a matter of speaking, we can take what is only partial—fuzzy or incomplete, in terms of holography—and put it into a bigger picture, thus sharpening our focus to realize potential in the explicate realm we know as the world. Remember that our inner systems are always unbalanced because of institutional, rigid thinking. The Great Restrictors are responsible for throwing our pictures of the universe out of focus, but with our Achievement Sciences, we can realign our lens and then bring focus to bear on whatever we choose. Having done so, we can let a "wave of consciousness" pass through our intelligences so that they may interact in true holographic fashion.

Perhaps a genius is nothing more than someone who has learned how to use his or her intelligences and realize potential. If so, then we all have the capability to far exceed our own imaginings, for thought and imaging are literally unbounded. It is scientific fact that humans only use three-tenths of their brainpower. Just think how different our lives could be if we started using even a fraction of the other seven-tenths of it!

Even something as simple as the act of driving a car demonstrates the holographic use of our intelligences. While driving, you are aware of hundreds of things at once and are continually processing incoming information without consciously thinking about it. You automatically deal with road signs, traffic laws, children playing in the street, the smell of exhaust, and many other things. Focus, of course, is where the activity begins. You are aware of getting into your car, your anticipated destination, the route you will take, and how long it will take to get there. You also conceptualize your car's Process, or the journey. In your mind's eye, you see the streets, the turns you will make, and your destination. You can be very creative because you are free to make new plans, such as taking an alternate route. Expression comes into play by actually driving the car. Procedure, Measure, Feedback, Parts, Synthesis, Physical, and Mental all play a part, too. To operate the car, you would need to

- Back out of the driveway and turn the wheel (Procedure)
- Gauge distance from other cars (Measure)
- Operate blinkers, gears, accelerator, brakes, etc. (Parts)
- Use your hands to perform the above (Physical)
- Use knowledge about driving, traffic laws, past experiences (Mental)
- Take into account the actions of other motorists (Feedback)
- Perform all of the above simultaneously (Synthesis)

The thirteen intelligences are used together in synergistic fashion, although some may not always be utilized for maximum potential at any given moment. They are parts of a symphony. There is no final or ultimate symphony, however, since there are endless ways to combine musical notes. This is also the case with human behavior. We are limited only by what we can dream or imagine.

The following information is an example of group intuition, wholeness and quantum theory. It addresses all of the quantum theories above. It works like this: Since systems seek balance, and potential seeks to be realized, we cross-discipline by taking various categories and/or components connecting them into a whole. The results will be a seemingly disconnected finding in our material world, but an actual glimpse of the quantum world that underlies our material reality. Throughout the book, we discuss quantum principles such as non-hierarchical status, coherent units, mechanics of the hologram, the Law of Complementarity, etc. Now we have a model that implements them all so as to demonstrate an underlying reality. This non-threatening model presents us with a new view of the issue and its details. In effect, it shows us a viable manner for the development and achievement of potential.

The following scenario relates to the manner in which the above process expresses its results. It was taken from a review of a new book, *The Wisdom of Crowds*, by James Surowiecki.

> At the annual west of England Fat Stock and Poultry Exhibition in the fall of 1906, a British scientist named Francis Galton became interested in a weight-judging competition: 800 fairgoers (a diverse group that included butchers, farmers, clerks, housewives, townspeople, smart people, dumb people, average people) tried to guess what a particular ox would weigh after having been slaughtered and dressed. The correct answer was exactly 1,198 lbs. After the judges awarded their prize, Galton borrowed all the entry tickets, did some arithmetic to get the mean of the fairgoers' guesses and found that their collective estimate was 1,197 lbs."[2]

The random gathering of people turned out to be an unexpected collective genius at ox-weight appraisal. Starting with this anecdote, James Surowiecki, financial columnist for the *New Yorker*, builds a fascinating case, summed up in his title and subtitle: *The Wisdom of Crowds: Why the Many Are Smarter Than the Few and How Collective Wisdom Shapes Business, Economies, Societies and Nation.*[3] Surowiecki's thesis posits an uncanny and generally unconscious collective intelligence working not by top-down dictate, but rather in dynamic arrangements of what the economist Friedrich Hayek called "spontaneous order."[4]

Surowiecki cites the giant flock of starlings evading a predatory hawk. From the outside, the cloud of birds seems to move in obedience to one mind. In fact, Surowiecki writes, each starling is acting on its own, following four simple rules: 1) stay as close to the middle as possible; 2) stay two to three body lengths away from your neighbor; 3) do not bump into any other starling; 4) if a hawk dives at you, get out of the way. The result is safety and a magical, organic coherence of motion—unconscious wisdom.

The old paradigm on this subject equates crowds with mindless mobs (the bigger the mob, the dumber and more dangerous). Think of lemmings or the Gadarene swine that Jesus sent off the cliff. The old paradigm, no doubt elitist and authoritarian, cherishes the brilliant individual (Leonardo da Vinci or Isaac Newton, who reinvented the universe while hiding from the plague in a country house).

The new paradigm, as formulated by Surowiccki, states that hoi polloi (the many) are weirdly smart and effective, even when a lot of them, as individuals, are average (or below) in their intelligence or their experience with the subject at hand. Surowiecki's sometimes Panglossian view sees a sort of invisible hand shaping the motions and outcomes of group phenomena (i.e., Shades of Adam Smith and the Theory of the Invisible Guiding Hand).[5]

Summary

While David Bohm investigated the holographic nature of the universe, Karl Pribram researched the holographic nature of the brain itself. Pribram found that memories are stored throughout brain tissue, with recollections remaining intact, though fuzzy, even when significant portions of the brain are removed. This corresponds with our previous model of the holographic picture of a horse, in which any part contains the whole.

We are not constrained, therefore, by linear modes of thinking dictated by conventional cause and effect logic. Drawing upon holographic principles, NATI proposes that we can realize potential as the result of whole brain thinking. The intelligences are therefore used in synergistic fashion.

By extension, we are then able to use complementarity. Using NATI principles, we can focus our intelligences, weaknesses, and polarities for the purposes of development. By first placing problems or goals into a "bigger picture," we can ultimately realize potential in the explicate realm we call the world. Whole brain thinking makes possible a higher level of functioning in every individual.

Chapter Thirteen

Correlating Traditional Science With Achievement Sciences

Buckminster Fuller called the Vector Equilibrium (VE) (Cubeoctaledron) a universal form of structure. Our model proposed by Derald Langham is one and the same! This natural geometric model is based on the mechanism of the VE in cellular development. These models are then interpreted as abstract symbols for our all inclusive model of understanding. We then demonstrate universality by integrating multiple theories of science with our 13 NATI Principles and Achievement Sciences. This integration and correlation demonstrate the all inclusive, comprehensive nature of the 13 Principles as basis for our Achievement Sciences.

While this chapter may not assist the reader in learning more about our subject matter, it will clearly demonstrate a direct connection of our natural systems theory to scientific doctrines. We think it is interesting and novel enough to include within the chapters. I invite the reader to seek out their own theory of science and find its correlation with NATI. As we have said throughout, the NATI system is based in science, with philosophical support. Each and every principle is explainable through scientific data.

Natural Thinking & Intelligence, while using philosophy and metaphysics to explain certain ideas, is based on scientific fact. Below is a list of the many scientific principles that correlate with NATI and its Achievement operating system for our everyday use.

1) Erwin Schrödinger's Cat Theory

Among other things, this corollary describes the duality of Polarity and its host. This is the famous theory that showed that two mutually exclusive outcomes can coexist in quantum mechanics. It postulates that a rather unfortunate cat is locked in a box that has a device capable of releasing lethal cyanide gas if a radioactive atom discharges. There is a fifty percent chance that the cat is alive at any given time if no one looks in the box, thus preventing any observation that would produce a definite outcome of the experiment. The Copenhagen School of Quantum Mechanics, built on the concept that the collapse of the wave function results from observation, cannot inform us as to the welfare of the cat.[1] In effect, this theory enables us to construct the NATI model so that we can deal with two complete opposites to formulate a position in our reality at the same time. We know that superposition of possible outcomes must exist simultaneously at a microscopic level because we can observe interference effects from these outcomes.

To understand Schrödinger's Cat Theory, one must accept how strange a theory quantum mechanics is. In all other scientific theories, we have models of how we think things work. For example, we know that distance traveled equals speed times velocity. If you travel for two hours at fifty miles an hour, you will go one hundred miles. Time is measured with a clock, while the distance is measured with the odometer on your car. Quantum mechanics is not like that, however.[2] What we measure in experiments is not described by quantum mechanics. Instead, quantum mechanics gives the probability that a given measurement or event will occur at some point in time. The only thing that quantum mechanics describes is how probabilities change with time. For example, if the particle in the cat example has a fifty percent probability of decaying in one hour, it will have only a slight chance of having decayed after one minute. After ten hours, it will have a high probability of having decayed. Quantum mechanics gives an exact model of how that probability changes over time. It says nothing at all about the actual state of the cat as these probabilities change. Science tells us what the probabilities are, but is completely silent on what (if anything) happens *between* observations.[3]

Schrödinger also developed the matrix as a form of quantum mechanics functionality. A matrix is a grid type structure or entity in which particles are embedded. In NATI terms, a matrix is a key part of the formulation of concepts, resolutions, and comprehension.

2) Neils Bohr's Principle of Complementarity

Here we see our Law of Complementarity, or Transcendence, as a basis of our system. This theory was adopted by Neils Bohr at the Copenhagen Convention of Quantum Theory in 1927. It was in response to the theory of wave-particle duality. In effect, complementarity says that bringing two opposites together forms a higher, third level of reality.4 One example is when colors are combined to form a complementary color. Another example is a complimentary duality between free will and determinism.

In NATI, this would be applied by looking for the integrating factor in any issue. Moreover, understanding the integration is a fundamental principle of intelligence and should always be pursued.

Complementarity in NATI also relates to the concept that a single intelligence may not be sufficient to explain all the observations made of various systems, so all thirteen intelligences must be considered when understanding a situation.

3) The Uncertainty Principle

Here we see that potential cannot be realized without clarity, definition, and awareness. Proposed by Werner Heisenberg, this theory states that we cannot universally know or measure all answers or all issues.5 Further, time and space cannot be measured simultaneously. Accordingly, energy and location cannot be determined at the exact same time. In NATI, a model must have the capability of changing without the model itself changing or else we will experience failure. Current behavioral and comprehension models do not utilize open and closed architecture, while NATI does.*

By utilizing the theory of probability and the notion of open and closed systems, we are able to identify multiple factors at the same time.

4) Chaos and Complexity

Here we will see the basis for building NATI Systems as well as interpreting patterns. In modern scientific terms, Chaos Theory was adapted or expanded by James Gleick (of the DNA discovery team).[6] It was advanced in the area of mind science by Ilya Prigogine.[7] Chaos, in essence, is recognized by many as patterns within other patterns or levels. These patterns are fixed and closed and operate independently. Reality is established by the visible parts of the different levels.** I have discovered a geometric example of the underlying nature of Chaos and how it is theoretically possible to comprehend it. With NATI's thirteen intelligences, we have the basis for decoding Chaos. The interaction of these fixed principles with variable factors generates a matrix and creates an infinite array of patterns. Chaos theory has been discussed earlier.

5) Polarity

Polarity, in simple terms, is oppositional forces such as positive and negative, white and black, good and evil, positive and negative. In NATI terms, Polarity exists in everything and is actually the driving force behind all understanding and behavioral factors. In order to gain understanding from a model, it must include both sides of an issue. This has been covered throughout the book.

---

*Other models (or understanding and reality) have little, if any, utilization of both open and closed systems.

**An example of this is Priorities. On one day our life focus may be on our work, while on another day it may be on an ailing relative.

6) Theory of Relativity

This is connected to Measure, Judgment. Developed by Albert Einstein in the early part of the twentieth century, this famous theory states that everything is relative within the universe and all things can therefore be measured in relative terms.[8] The first rule of relativity states that the laws of physics must be the same for observers in different frames of reference. While observation and measurement are relative, there are absolutes such as the speed of light. In NATI terms, observation and measurement are relative not only to each and every other non-absolute thing, but also relative to their own frames of reference. Every event or notion can then be traced to its invariant source, which is one of the thirteen forms of intelligence and/or Polarity.*

Everything else is then variable and is implemented in the process of reality.** In NATI, we apply this principle to our model in order to demonstrate how opposites and unrelated factors do indeed connect. Further, things must be compared to a core focus, and these things require consistent measurement.

7) Theory of Laws

The Principle of Models, Rules and Laws is based here. This ancient theory provides us with boundary, structure, and rules by which participants behave. This is one of our thirteen intelligences and creates a model that we follow to bring us insight.

8) Holographic Theory

The Concept of Wholeness, Integration and Synthesis is found here. This twentieth century finding, which was promoted by Karl Pribram[9] and David Bohm[10], was based upon the discovery of the holographic picture. According to this phenomenon, the

---

*Remember, we are stating that the thirteen principles/intelligences are basic to our nature.

**Everything else covers such factors as Cultural Characteristics, the Great Restrictors, Core Human Dynamics, etc. It is these factors that generate the relativity in our lives, while the thirteen are fixed. It is our interpretation of them that shapes our reality.

parts of the whole are a representation of the whole and vice versa. We implement Holographic Theory throughout the book by connecting exclusive and unrelated factors.

9) Human Factor Formula

This is an adaptation of the ancient philosophical premise first developed by Thales, the ancient Greek scientist and his famous student, Pythagoras. This we identify in the book as the formula:

A + B = C

A = Awareness B = Beliefs C = Communication

10) The Looking Glass Theory (Visceral Science): Using The Mirror

This is a valuable notion for an accurate and exciting theory. I utilize this term because of the definitive insight that it gives one in comprehending his or her own processes. *Looking Glass Universe*[11] makes a profound argument for the fact that the universe is indeed a mirror of itself; in the same fashion, so are we. This is one of the most effective and profound principles in NATI and in nature.

11) Science of Functionality

This is a centuries-old basis for the way we take action. This theory has its basis in numerous ancient societies, including China, Greece, and Hebraic cultures. It was described as fire, earth, air, and water and represented the elements from which the universe and humankind were created. In NATI, we utilize this to identify the four ways by which things function physically, emotionally, mentally, or intuitively.

12) Multidimensionality

This is where Details, Parts and Cross-Disciplining of Systems are based. This twentieth century notion deals with the laws of various levels of reality and their differences from each other. It originated from life-science and uses methods and technology from other disciplines. Ludwig von Bertalanffy is one of the key creators of General Systems Theory and multidimensionality. Within the Internal Archives of Bioscience Organization this theory is also quite prevalent. We use this concept not only to connect the various distinctive notions of a situation and reality, but also to understand how various levels and parts have their own existence while they simultaneously move within a greater community. A classic example would be the human body and its cells. In NATI, the thirteen intelligences are individualistic, yet act collectively to form varying multiple systems, or dimensions. NATI makes it easier to define and locate systems, patterns, and procedures.

13) Natural Laws

This is another ancient concept that serves as the basis of our findings, for this is where the Rules, Models, and Laws come from. This has been defined throughout the ages as the order that governs the universe. Thomas Aquinas said eternal life is God's wisdom inasmuch as it is the directive norm of all movement and action. God directed us to an end. The role that God has prescribed for our conduct is found in our nature. In NATI terms, and in terms of General Systems Theory, we take that a step farther and say this role is not only found *in* our own nature, but also *throughout* nature. Those roles in human terms are found in the thirteen principles plus Polarity. We further conclude in NATI philosophy that God and nature are intertwined and that nature defines the human mind. Moreover, this nature emanates from potential and represents an absolute intelligence of the universe.

14) Theory of Potential

This important theory actually implies an impersonal description of God. Potential is identified as the substance of the absolute, the God-force. Development is its (and our) objective. Its basic elements, as found in Chaos Theory, are creativity and Polarity. In NATI terms, we have further identified the segments of potential as being the Creative, the Organizational, and the Functional. Potential, in itself, has been defined as "something that does not exist, but which can be imagined. It is latent." This theory opens up an entire world of creativity and understanding insofar as it enables the human mind to accomplish anything that it envisions without question.*

Potential energy is the energy stored in a body or system because of its position, shape, or state. In NATI, it has a direct bearing on frame of reference. Further, there is a principle called plenitude. This states that the universe, to be as perfect as possible, must be as full as possible; it must contain the greatest possible diversity (and profusion compatible with) of kinds compatible with the laws of nature. This principle certainly supports the concept of potential as a universal force rather than something that is strictly an idea because it is all encompassing. Every possibility exists within it. Some call it the mind of God.

15) Open and Closed Systems

Not only do we have Polarity here, but Order, Process, and Development as well. These two factors are a result of modern-day systems sciences and represent the nature of Concepts and Models in NATI terms.** All open systems create reality. They are also vital for depicting random results from normal patterns and finding order out of randomness. Open systems represent unstructured models, while closed systems represent structured

*If it can be imagined with full detail and believed in, it can be accomplished.

**They represent the structure of our thirteen intelligences in that we plan, organize, and function in open and/or closed modalities. For example, planning something with a closed mind is not going to produce as good a possibility as if it were an open mind.

models. In NATI, what we have created by our model is a closed system with certain attributes of an open system, thereby enabling unbridled results.[12]

16) The Theory of Probability

Here is another interesting notion for Measure, as well as how NATI works for each of us. This is a branch of math concerned with numbers that are called probabilities and reflect the concept of chance. Probabilities occur all the time in science because we almost never know everything we need to make a completely accurate prediction. For example, if you really want to make a trip of a hundred miles, you cannot know ahead of time exactly how long it will take. You might run into a traffic jam. You can only give an estimated time.

In quantum mechanics, probabilities are different. They do not result from our limited understanding of the mechanics of the universe. The theory states that an event that occurs a certain number of times, called "successes" in similar tries, has a probability that is the ratio of the successes to the number of tries. In NATI terms, we say that probability is evidenced by the fact that eventually one of the thirteen intelligences and/or Polarity will bring us to the answers we are seeking. In virtually thousands of applications involving numerous types of clients and situations, this system has responded favorably without question.*

17) General Systems Theory

Here we have the model for the transition from nature to Human Understanding and its systems. Ludwig von Bertalanffy,[13] a key creator of General Systems Theory (GST), was one of the most important theoretical biologists of the first half of the twentieth century. He conducted research on comparative physiology, biophysics, cancer, psychology, and the philosophy of science. He developed a kinetic theory of stationary open systems, was one of

*The source here is the Gilchrist Institute and the author's twenty-five years experience.

the founding fathers of the Society for General Systems Theory, and was one of the first to apply GST methodology to psychology and the social sciences.

By the 1930's, von Bertalanffy formulated the Organismic System Theory, which later became the kernel of General Systems Theory. His starting point was to deduce the phenomena of life from a spontaneous grouping of system forces comparable to the system developmental biology of today.*

He based his approach on the phenomenal assumption that there exists a dynamic process inside the organic system. He postulated two biological principles, namely the maintenance of the organism in non-equilibrium, and the hierarchical organization of a systemic structure. As a metatheory derived from both theories, von Bertalanffy introduced GST as a new paradigm that could control the model construction in all the sciences. As opposed to mathematical systems theory, it describes its models in a qualitative and non-formalized language. Thus, its task was a very broad one: to deduce the universal principles that are valid for systems in general.

He began by reformulating the classical concept of a system and defined it as the relation between objects and phenomena. As a methodology applicable to all sciences, GST encompasses the cybernetic theory of feedback that represents a special class of self-regulating systems. According to von Bertalanffy, there exists a fundamental difference between GST and cybernetics since feedback mechanisms are controlled by constraints, while dynamic systems show the free interplay of forces. Moreover, the regulative mechanisms of cybernetic machines are based on predetermined structures. In short, GST is a regulative instruction that synthesizes the data, or even laws, of the natural sciences and makes them applicable to all the other sciences. By this, a matrix model is formulated so that data can be categorized, analyzed, and synthesized.

---

*In other words, he applied the observation of various biological systems working independently as part of a collective group.

This theory not only represents the organizational principles of Feedback, Models, Details, and Holism, but also clarifies the system of self-regulation (i.e., how we establish procedures and programs that can guide and correct us through a feedback process).

18) Continuum

This is another theory of Parts and Details, as well as abstract and concrete thinking. Interestingly, Continuums work well with abstract thinking. With an abstract view of an issue, a spectrum of related possibilities is presented, while with concrete thinking, a very specific view is presented. In science, continuum is a set of identifying points on a line. In NATI, it represents a number of principle forms. These forms are sometimes called archetypes. There is any number of synonyms or related interpretations for a particular NATI principle. An intelligence continuum is a group of terms that functions in conjunction with a frame of reference. We use an archetype as a potential meaning in place of the original idea or intelligence. In this respect, the various intelligences actually are represented as a group, or principle. This means that we are looking at symbols, not necessarily specifics (e.g., a Buick symbolizes a car). We have seen such archetypes in the various lists and charts of the thirteen intelligences and their synonyms.

19) Coherent Units

This is how we can achieve a resulting system or data model. In science, this is defined as a system of measurement in which units are obtained by multiplying or dividing base units without the use of numerical factors. In NATI, this is analogous to the matrix of the thirteen intelligences subdivided into various information subsets, such as with the various potential decoder matrices. For instance, when we create a matrix of Core Human Dynamics vs. A + B = C, we then have an expanded model of analysis.

20) Empirical

With over 2000 cases and analyses in our records, we can safely confirm the veracity of the NATI program. This denotes a result that is obtained by experiment or observation rather than from theory. While we have theory in NATI, we do have a very strong sense of empirical formulas by virtue of NATI's interactive feedback structure. The very organization of the thirteen intelligences themselves generates an empirical formula for the user. Predictable results (i.e. problem solving) evolve through implications of the system.

21) Feedback

The best scientific proof of the Mirror is that it works! Feedback is the use of a system's output to control or direct its performance. In *positive* feedback, the output is used to enhance (or increase) the input. In negative feedback, the output is used to eliminate (or reduce) the input. There are numerous other uses as well.

22) Self-Organization

Nature & Absolutes such as virtue all possess this quality, as do our NATI Principles. This is the spontaneous order arising within a system when certain parameters of the system reach critical values. It is related to the concepts of broken symmetry, complexity, and non-equilibrium mechanics. Many systems that undergo transition to self-organization also undergo transition to Chaos. Typically, what may appear to be confusion and Chaos is nothing more than nature using a path that we are unfamiliar with to obtain its objective. One difficulty people often have lies in their inclination to follow a path blindly, without knowing or seeing what is coming next. That is why self-organization is not easily accepted.

The basic principles of NATI enable self-organization. Not all closed systems foster self-organization or imply procedures.

NATI systems enable flexibility where most other systems do not. They are simply too inflexible.

Summary

Although our Achievement Sciences have a philosophical underpinning of metaphysics and philosophy, they correlate with hard science. The twenty-two laws and theorems listed in this chapter all describe principles that are reflected in nature and validated by one or more branches of science.

Chapter Fourteen

Core Human Dynamics and Life Philosophy as Disruptive Systems

A Cautionary Tale

What is it that disrupts our human systems? Let's take a look at some prospects. Scientists tell us (and Ilya Prigogine, in particular) that dynamic, open systems are healthy and maintain equilibrium. We have also noted that our inner matrices and COPS are invariably corrupted by the Great Restrictors since we are subjected to inflexible institutions (which by the way can include the family) from the time we can walk and talk. In some cases, the resulting closed systems can have devastating consequences.

A classic example of this is what we might term "radical patriots" such as Timothy McVeigh, the man who was indicted and executed for the Oklahoma City bombing a few years ago. McVeigh, as you might recall, claimed that he was acting out of patriotism, protesting what he regarded as the infringement by the federal government on the rights of individuals. He was particularly upset by the attempt of federal agents to inspect two areas suspected of housing caches of illegal weapons: Randy Weaver's home in Ruby Ridge, Idaho, and David Koresh's Branch Davidian compound in Waco, Texas. In a more general sense, McVeigh was upset by the federal government's suspicion of white supremacist groups and non-governmental state militias. McVeigh's action in Oklahoma City resulted from a closed system that regarded the U.S. Constitution literally, not as the flexible document its framers intended it to be. The Constitution, specifically, was intended to be an open system, capable of interpreting rights under circumstances of constant flux, with a subsystem of amendments available to change the document itself when warranted. The Constitution is an excellent example of both Order and Laws (among other things). As we have seen, these two intelligences are fixed and do not change, but how people use them is most assuredly a matter of variability. The failure to recognize this dynamic, variable nature of the Constitution was (and is) a basic flaw in the systems of radical patriots.

By the same token, it can be said that the federal government's attitude toward those with overly sensitive systems, such as McVeigh, was also indicative of a corrupted matrix. It could also be argued that the government's failure to recognize and deal with the radical patriot mindset eventually brought about the bombing. This is not to defend McVeigh's action in any way, of course. While the U.S. government may be far from perfect in dealing with people like McVeigh, patriots cannot effect positive change without having open, living belief systems.

The same case might be made relative to terrorism. The systems of terrorists are closed, inflexible, and are based on all of the Great Restrictors—Fear, Ego, Ignorance, and Self-deception. The purity of their intent is always sullied. In their minds, evil and goodness cannot be approached as anything but irreconcilable opposites. They cannot find complementarity in their thinking as they attempt to demonize various governments or religious factions. As in the case with the Oklahoma City bombing, it may be argued that the government should be far more aware of the terrorist mindset and use different strategies to deal with systems so far out of balance. I am not making a political statement here, but am confining myself to what mind sciences say about open and closed systems and the best way to handle a lack of equilibrium. In effect, radicals are one end of the Polarity spectrum. *As long as society rejects the fact that radicalism is a valid component of nature, we, as a society, will continue to suffer its consequences.*

Open and Whole

We will make sure that our own discussion is balanced by now looking at healthy approaches to systems. A book published by the Institute for Personnel and Development talks about how to facilitate learning within companies.[1] The facilitative approach, as it is called, helps people to "learn how to learn" and motivates them by giving them responsibility for their own development. The book states that if corporations are to survive, they need to continually question the way they work, determine their customers' needs, and provide a high quality of service. Furthermore, they need employees who relish change and take every opportunity to be creative and innovative, people who actively seek responsibility. These kinds of employees need to be energized, not directed. The facilitative method is a "whole systems" approach to learning and application because it is centered around factors such as creativity, development, and feedback. As our own research at the Gilchrist Institute has shown, giving people more responsibility for their development is crucial to their growth and eventually reflects the well-being of both society and the companies they work for.

Organizational Networks

A recent article by a senior consultant at Ernst and Young concerned organizational networks in the business world.[2] This article discussed what are, in effect, principles of both Systems Theory and Natural Thinking & Intelligence.

The first issue the article dealt with is what is called Multidisciplinary Focus and defined a network as a multi-disciplinary group of managers chosen by the CEO. Ideally, the CEO is qualified by good judgment, motivation and drive, and control of resources at junctures of critical information flow. Such people are uniquely qualified to determine a firm's corporate strategy. The network, in general, reshapes the way business decisions are made and by whom, integrating decision-making horizontally at various managerial levels.

Furthermore, networks are omni-competent, redirecting the flow of information, decisions, power, and sources of feedback. This high level of competence helps shape personal relationships within the organization and allows other managers and members of the network to make decisions, too. Because this dynamic system is structured horizontally rather than vertically, members can be constructive in their approach to conflict, creating solutions rather than pointing fingers. Sound judgments can be made, helping people to feel unique, accepted, and appreciated.

In short, structuring an organizational network on this kind of multi-disciplinary ethic produces a sustained focus on the fundamentals of the business instead of on culture, ego, or politics. This has a tremendous effect on both the company and society as a whole.

Core Human Dynamics: Basis For Life Philosophy

In the organizational network just described, we noted that company employees possessed characteristics such as drive, motivation, acceptance, uniqueness, and good judgment. NATI believes that characteristics such as these are fundamental to the human condition. There are quite a few, but Achievement Sciences identifies the following as the most important dynamics responsible for human behavior. They are:

- control (influence or regulation)
- power (potency, strength, or energy)
- inclusion/acceptance (being a part of something or containing other parts)
- exclusion/uniqueness (restrictive or reluctant to accept)
- attention/recognition-seeking (calling attention to oneself or being notable)
- self-interest (interest in personal advantage, motives, or comfort)

- judgment/values/appraising (opinions, valuations, conclusions)
- drive/motivation (to act, instigate, or desire an action)

Core Human Dynamics (CHDs) should not be interpreted as either positive or negative, but as neutral. In any given situation, they may produce healthy or unhealthy behavior. Timothy McVeigh, for example, certainly felt that people who shared his beliefs were being excluded from society. He therefore made certain appraisals of situations (events at Ruby Ridge and Waco) and then sought control through power. As we can see in the above corporate model, however, these very same core dynamics can be used in very positive, productive ways. My own experience is that people have an over-abundance of one or more of these in their personality.

Core Human Dynamics combine with the basic 13 principles to form recognizable mindsets. For example, if one focuses on a Concept or Belief concerning control, he or she will naturally and automatically consider the following:

- rules about control
- priorities regarding control
- integrating control
- the measure of control
- the process of control
- details of control

The same process would naturally apply to other Core Human Dynamics:

- rules about power, exclusivity, judgment, etc.
- priorities regarding power, exclusivity, judgment, etc.
- integrating power, exclusivity, judgment, etc.
- the measure of power, exclusivity, judgment, etc.
- the process of power, exclusivity, judgment, etc.
- details of power, exclusivity, judgment, etc.

Moreover, by combining the CHDs with the Functional categories, a more in-depth understanding of people's mindsets and actions is revealed by considering their:

- rules concerning physical, mental, emotional, or intuitive control
- physical, mental, emotional, or intuitive priorities |regarding control
- physical, mental, emotional, or intuitive integration of control
- physical, mental, emotional, or intuitive measure of control
- physical, mental, emotional, or intuitive processes of control
- physical, mental, emotional, or intuitive details of control

By following the above models of the various NATI principles, we can readily ascertain an accurate life philosophy. Core Human Dynamics are very important because potential is realized through your mindset. Your mindset is what you are—the very seat of your soul—and what you are at present naturally determines what you can become, as well as your overall purpose in life.

In hindsight, we certainly know what kind of rules, priorities, beliefs, and patterns of control and power dominated the mindset of Timothy McVeigh. His radical patriot mindset, hindered by the Great Restrictors, ultimately led to certain mental, emotional, and physical functions relative to CHDs. Productive corporate strategies, using the same CHDs, lead to entirely different functioning. The former mindset represents the quintessential closed system. The latter represents the open, continually evolving system.

Blocks of Potential

There are several key blocks to our potential. Basically, the Great Restrictors we have outlined are the most prominent.

Preconceived notions, like those of McVeigh, are further blocks since they cause visualizations of the way things are. Emotionalism—that is, reacting—is yet another. Connected to all of the above are cultural characteristics. The human mind buys into a group's view of reality (just as McVeigh bought into the mindset of various state militias). Religious, societal, corporate, and governmental institutions are prominent examples here.

For the sake of an objective discussion, let's deal with the corporate culture issue. The bottom line is that I have witnessed numerous examples of corporate culture severely restricting corporate growth. For example, new business models are discovered, and the first thing corporations do is to try to convert them to their own modality! I have a good news—bad news example utilizing a "spirit in business" model. The good news is business has discovered spirit, while the bad news is that it is reengineering it and negating its potential.

These Cultural Characteristics are keys to our life philosophy. We become restricted to acting and thinking in ways other than what is presented to us. They become our values. This is why virtue is superior as a path to values. With virtue there is no restriction, just direction.

Summary

We can see examples of closed systems in radical patriots such as Timothy McVeigh, convicted and executed for the Oklahoma City bombing. His inner matrix was closed, with Ignorance preventing him from understanding the true flexibility of the Constitution. His life philosophy was extremely limited and eventually led to a breakdown of his system.

There are many examples of healthy, thriving systems, however, as seen in management styles that allow people to assume responsibility for their growth in the corporate world.

Core Human Dynamics are the major characteristics of the human condition. They are interpreted as neither positive nor negative, but neutral. They are:

- control
- power
- inclusion/acceptance
- exclusion/uniqueness
- attention/recognition
- self-interest
- judgment/values/appraising
- drive/motivation

The closer we come to comprehending systems of nature and their components, such as Polarity and Wholeness, the greater our ability to achieve our potential. We must be careful not to compromise our drive toward realizing that potential, however, by falling prey to the Great Restrictors, preconceived notions, or emotionalism. Additionally, we must be careful to follow the path of virtue rather than values presented by group mentalities.

Chapter Fifteen

How Heroes apply NATI Intelligences: From Lebanon to The Moon

"Houston, we have a problem."
—Jim Lovell, Commander of Apollo 13

Prominent figures in history have accomplished great tasks and produced marvels that are admired by nearly everyone in society. We have described the intelligences they used, intelligences that we *all* possess. In this chapter, we shall begin a process of synthesis that, in one form or another, will continue throughout the rest of this book and show that these intelligences permeate everything we do. They are not isolated or compartmentalized within the human organism, but constantly interact with each other in dynamic, ever-changing patterns.

Below are all the intelligences we have discussed, although it is more accurate to think of them collectively, as a matrix, rather than as a sequential list.

Creative
- Focus/Awareness
- Beliefs/Concepts
- Communication/Expression

Organizational
- Laws/Models
- Parts/Details
- Order/Processes

- Measure/Assessment
- Reflection/Mirroring
- Wholeness/Synthesis

FUNCTIONAL

- Physical
- Emotional
- Mental
- Intuitive/Spiritual

TERRY ANDERSON

In order to see the NATI intelligences working together, we shall look at two extended examples that represent a more holistic model of the system. I want to begin with the ordeal of my friend Terry Anderson, a journalist who was abducted by Islamic militants in 1985 and held hostage for almost seven years in Beirut, Lebanon. Terry's plight was one that few people will ever have to face, but his survival shows that NATI is a dynamic system, one that can be used to solve even very serious problems by the application of the thirteen intelligences we have defined. Let's examine how each one was used by Terry during his captivity.

Focus/Awareness: Terry's *lack* of awareness was surely his downfall. He did not heed a kidnapping attempt the day before he was actually abducted, and he has stated in many interviews that, given the inflammatory nature of the Middle East, he was not as careful as he should have been. Once he became a hostage, however, his focus shifted from wants to needs, for survival most definitely did not include luxury items. He also accepted the fact that he might be confined for a long time and refocused his attention on his new, although quite undesirable, environment. Courses in survival training always teach people to focus on the immediate situation and tasks at hand. Panicking or continually asking "why me" not only squanders precious energy, but it doesn't enable the brain to concentrate on the elements that require attention if one is to stay alive.

Beliefs/Concepts: Terry imagined himself to be a hostage, not a journalist. He accepted his fate and the fact that he was going to be living in a different culture. He knew the grim reality of his situation. He also realized that, in the short term, he was powerless to obtain his freedom. His mind came to grips with all of the unpleasantness associated with the concept of being a hostage. For Terry, assuming the identity of a hostage and believing in the reality of captivity was no different than someone believing himself to be a good citizen, educator, Muslim, Christian, or politician. It was who he was.

Communication/Expression: He therefore became a professional hostage and acted accordingly. His words and actions were commensurate with those of someone who is a captive. He could be very open or very silent, depending on what level of cooperation he wished to display. A wide range of social communication was no longer available to him on a regular basis, so he chose his modes of expression carefully. He needed to display both caution and presence of mind in communicating with his captors and fellow hostages. (After being released, he found his reintegration into society relatively easy because he had had almost seven years of practice in focusing his awareness on the manner of his expression. He used his recognition-factor beneficially and communicated his ordeal in his bestselling book *Den of Lions.*)[1]

Models/Laws: Although Terry followed the day-to-day rules prescribed by his captors, he also followed the laws of survival. His objective switched from the more immediate "I have to get out of here!" to "I am being detained against my will and I must stay alive at all costs." His new focus on the laws of survival was quite similar to that which people have when lost in the wilderness. Extreme necessity gives us many new Models that work with Priorities to show us what Procedures to implement.

Patterns/Processes: Terry adopted positive Procedures. He asked for better conditions for himself and his fellow hostages as a way of implementing the laws of survival. At times, he *insisted* that certain critical needs be met. Other times, he was willing to negotiate. In yet other situations, he was completely passive since he realized that some conditions were unchangeable. In short, he did what needed to be done as opposed to what he wanted to do.

Mirroring/Feedback: As we have already seen in a previous chapter, Terry learned about his own behavior from the unpleasant exchange with another prisoner. He found that he himself was sometimes obstinate and inflexible (when such obstinacy was clearly uncalled for). He therefore learned from his experiences, both positive and negative. He was willing to experiment and try different methods of survival based on the feedback from both his captors and other hostages. It took a great deal of courage to look at his own shortcomings when the most obvious inclination was to focus only on the unfairness of his situation.

Assessment/Priorities: He decided what was important and how to achieve it. He established his dignity as a human being and stuck to it no matter what. When he wanted to communicate with his family, he continually battered his head against the wall literally—even to the point of drawing blood—until his captors relented. He paid close attention to the behavior of the militants so he would know when a straightforward request was likely to be honored, but he always assessed what it might take to effect change if those in charge proved intransigent on any given matter.

Parts/Details: Terry tried to make the most of everything, utilizing whatever resources were at his disposal. This is critical in any survival scenario. In a hostage situation, one makes use of simple things, such as the permission to move or speak, even if speech and movement is limited. A sip of water, a piece of bread— these are never disdained but are regarded as part of the larger picture of staying alive. Terry always tried to generate more resources, such as medicine, food, water, and exercise, as well as mental resources, such as the gratification that came from seeing that he could make himself heard. Perceiving success is an important aspect in the step-by-step process of staying alive in a precarious situation.

Wholeness/Synthesis: He brought everyone into the experience of being a hostage, even the guards. When he went into the bathroom and saw a machine gun leaning against the wall, he walked out and told the guard, smiling, that there was an unattended weapon in the bathroom. In short, he used physical, mental, emotional, and spiritual components to survive, and he worked with everyone he was with to achieve desired results. He learned that his captivity was a synthesis of East and West in terms of politics, goals, and behavior.

A Successful Failure

Apollo 13, launched on April 11, 1970, is a different kind of survival story, although it shares many elements of Terry's struggle in Lebanon. Two previous missions, Apollo 11 and Apollo 12, had allowed men to walk on another planetary body for the first time in human history. Manned flights by American astronauts had always been successful, and the number thirteen (usually considered unlucky) was naturally ignored. The flight was launched at 1:13 pm Houston time, and Apollo 13 entered the moon's gravitational field on April 13. NASA was not an agency that entertained superstition. It was a task-oriented group of men and women who lived and breathed science, discipline, and logic.

As we now know, the flight proved to be one of the most dangerous in the entire Apollo program. Jim Lovell, Fred Haise, and Jack Swigert, Jr. were nearly 200,000 miles from earth when Oxygen Tank #2 exploded, crippling the service module that carried air, water, and power for the command module, named Odyssey. The explosion left the Odyssey with only ten hours of reserve life-support that needed to be saved for re-entry, assuming a plan could be devised for returning the astronauts to earth in the first place. They were, after all, headed in the wrong direction in a crippled ship. The crew's struggle to survive illustrates superbly the use of all thirteen intelligences (as well as the intelligences of their support team at Mission Control in Houston).

First, let's examine the environment in which the astronauts had to live and work, namely the spacecraft. It consisted of three distinct sections that were part of an overall system designed to send them to the moon and enable them to land near the Frau Mauro mountain range. The command module, Odyssey, was the Apollo capsule in which Lovell, Haise, and Swigert were based for both lift-off and splashdown. It was the central component of the spacecraft, where the main control systems and instrumentation were located. The second component of the spacecraft was the service module, which was a long cylinder attached to the rear of the command module. As noted, it was the command module's power plant. The last component of the spacecraft was the LEM, or Lunar Excursion Module (named Aquarius), which was to separate from the command module while in lunar orbit and land on the moon. The importance of Parts was obvious. It was only after a handful of scientists lobbied NASA for years that the agency finally admitted that a modular system was required to take man to the moon. The "single vehicle" theory called for a massive amount of fuel that would weigh almost as much as the rocket itself.

While the service module endangered the crew's lives, the system as a whole saved them in the end, for with very limited power at their disposal in the Odyssey, the astronauts used the third part of their craft, Aquarius, as a lifeboat during the eighty-six hour return trip to earth. The LEM's use as a lifeboat was a clear application of Models. On a large ocean-going vessel, a lifeboat is a small craft with only the attributes of seating and seaworthiness. In the same way, Aquarius, intended to function as a lunar lander, was not designed to function as crew quarters during the journey. Nevertheless, it had seating capacity and was "space-worthy." The connection of these three parts, therefore, quite literally saved the lives of the crew. (Notice that Parts and Models were operating simultaneously and synergistically.)

The need for a lifeboat model, of course, stemmed from an initial awareness of the service module's problem. Engaging in a practical joke before the oxygen tank exploded, Fred Haise opened a valve, producing a loud *bang*. A few minutes later, another bang was heard by Lovell and Swigert and attributed to Haise's fondness for jokes. It wasn't until Swigert noticed that a master alarm had sounded, signaling a power drain, that the astronauts began to suspect that something was seriously wrong. Even then, however, the magnitude of the problem was not fully perceived. The crew did not yet believe their lunar landing was in jeopardy. It wasn't until Commander Lovell looked outside the cabin window of the Odyssey and saw that the service module was venting precious oxygen into space that he became completely aware that the mission objective was to survive rather than to land on the moon. The astronauts were trained to be calm under pressure, and it literally took the above series of events to finally focus their attention on the seriousness of their predicament. Their first reaction was analogous to Terry Anderson not taking seriously the preliminary kidnapping attempt by the Islamic militants. In both scenarios, there was an initial, though brief, lack of Awareness.

In their desire to get back home, they then implemented their Emotional intelligence. Immediately, the crew began making critical Assessments. How badly was the service module damaged? Since the entire spacecraft—all three components—was pitching wildly through space, how could they regain proper attitude and alignment? What was left in the way of resources—food, water, oxygen, and power? How could they put the craft on a return trajectory for earth? Was return and reentry even possible given that so many essential systems had been lost or seriously compromised? Could the Aquarius indeed sustain them for over three days?

With little time to spare, the crew and their ground support rapidly brought considerable brainpower—their Mental intelligence—to bear on the problem, using every bit of knowledge they had as they considered scientific Laws that had to be taken into account, laws dealing with gravity, propulsion, motion, and human biology. Without thorough knowledge of scientific laws and principles, all other actions would have been utterly meaningless.

NASA personnel were naturally knowledgeable in matters of science and engineering, so their knowledge was used to first prioritize and then adopt effective Procedures. Assessment and Process, working in tandem, were now being directly applied to the crisis.

Prioritization took many forms. Immediate survival took precedence over any other consideration, so the crew was moved into the LEM within thirty minutes of the explosion. Next, calculations were made to determine how much power, oxygen, and water would be available for the crew. All the technical capabilities in the world (or in space) would be useless unless the astronauts could be kept alive with provisions necessary for life support. Their physical bodies needed attention if they were to continue "working the problem," to use NASA terminology. It was also deemed critical to shut down all remaining electrical systems aboard Odyssey to preserve what little power it had since a small amount of electricity was required for successful reentry into earth's atmosphere. All of these operations *had* to be performed before any others.

Procedural intelligence then shifted to carrying out various tasks. Checklists had to be consulted. Valves and switches had to be turned off as the crew moved from Odyssey to Aquarius. Aboard Aquarius, power-up procedures were necessary to make the LEM habitable. On the ground, a complex network of computers, engineers, and designers was assembled to provide information on the systems in all three modules, consulting blueprints and specifications that were normally taken for granted during each flight. These individuals, with their computers and simulators, were part of a grand scheme of Parts and Details that was, for all practical purposes, the amalgamation called "the space program." As they networked and communicated their knowledge, however, they were transformed into a single, functioning unit.

Thus, Synthesis and Expression were now working with Parts, Models, Focus, and Mental.

In the matter of communication, the movie *Apollo 13*, based on Jim Lovell's book about the flight (*Lost Moon*), showed accurately how communications on the ground and in space were sometimes quite "ragged" as the personnel and flight crew grew a little punchy.[2] Throughout the ordeal, though, the character of their Communication remained, on balance, positive and professional. Here we see the truth demonstrated that A + B = C. They had extreme focus on their goal, plus everyone ardently believed that it was possible to get the Odyssey home safely. As flight director Gene Kranz told his subordinates, "Failure is not an option." The ability of NASA personnel to communicate large volumes of data from coast to coast quickly and efficiently was a direct function of their Awareness of the problem and their Belief that success was possible.

The scientists at NASA decided to bring Apollo 13 home using what was called a "free-return trajectory," meaning that the moon's gravity would slingshot the spacecraft around the moon and send it back to earth. To accomplish this, the LEM's engine was to be used to alter the course of Apollo 13. Later, the engine of the Aquarius would also be needed to make a course correction since the spacecraft was not predicted to hit the earth's atmosphere at the proper angle. If it entered too steeply, it would burn up upon re-entry. If the angle was too shallow, it would skip off the earth's upper atmosphere and travel back into space with no ability to turn back or make a second try since the command module had no engines of its own. But here was the rub: the LEM engine was not designed to act as the main propulsion unit for all three component vehicles. It was designed for lunar descent and lunar liftoff only—nothing more. Nevertheless, Mental and Intuitive intelligence told NASA personal that the idea could work. To test this application of the LEM's engine, astronauts from previous missions worked in ground-based LEM simulators to see if the proposed "burns" by Aquarius would be successful. Without the intelligence of Models, it is doubtful that the correct calculations—when to ignite the LEM engine and how long to burn it—could have been made.

Although the Aquarius' engine performed successfully, the astronauts were "not home yet," figuratively or literally. The crew's endurance and health would now be tested, bringing Physical intelligence into play. There was no internal heat source in Aquarius or Odyssey, so that cabin temperatures hovered at thirty-eight degrees Fahrenheit. This hampered the crew's ability to sleep, caused fatigue, and produced erratic emotions, which as mentioned earlier, partly compromised the quality of their communications with Houston and with each other. Haise also developed a urinary tract infection because water was rationed and he was not able to drink enough fluids. In this extremely cold environment, he ran a fever of one hundred and four.

Apollo 13 continued to encounter obstacles. It was discovered on the way back to earth that carbon dioxide levels in the spacecraft were becoming dangerously high. Human respiration involves the intake of oxygen and the exhalation of carbon dioxide, which can become lethal if inhaled in large quantities or in a confined space. In the frenzy of activity to accomplish the free-return trajectory, engineers had overlooked the fact that Aquarius was designed to hold two men for approximately two days while on the moon. Three men were now using the close quarters of Aquarius for a time period approaching three days. The carbon dioxide filter in the LEM simply couldn't handle the load.

A jury-rigged filter was fashioned from available materials in the craft: duct tape, cellophane, metal canisters, and cardboard covers from flight plan booklets. Using Models and Intuition, the ground crew converted Parts into a Whole, always driven by the desire that is part of Emotional intelligence.

The new CO_2 scrubber worked, but as the three modules approached earth, NASA didn't know how to provide Odyssey with enough amps to power up its systems again. Sufficient electricity was necessary to jettison the wounded service module, separate from Aquarius, re-activate the main computers, and fire small rockets, called "pyros," that would help deploy parachutes to slow the craft as it fell toward the surface of the Pacific Ocean. Working in a simulator, astronaut Ken Mattingly experimented with transferring Aquarius' remaining power to Odyssey. Physical intelligence was involved in the transfer of energy from one set of batteries to another. Indeed, throughout the entire mission, we see that the Physical continuum of matter and energy was always in play as the crew tried to both conserve energy (electricity), as well as transfer it from one module to another. We also shouldn't forget that space flight was (and is) made possible by the conversion of matter (fuel) into the kinetic energy of motion.

The crew did indeed make it safely back to earth, and their journey is now legendary in the annals of manned space flight. One factor that cannot be ignored is the high level of Intuitive and Spiritual functioning that enabled Apollo 13 to become the most "successful failure" of the National Aeronautics and Space Administration. The astronauts themselves went beyond normal limits of physical, emotional, and mental endurance. Even long years of training had not prepared them for such an ordeal since most of the procedures used to bring them home had to be invented during the tense hours of the crisis. This called for the crew (and the connected ground network) to function beyond their normal abilities. It may be said that their intelligences were kicked into overdrive and operated synergistically to convert potential into survival. Hunches and instincts reached spiritual dimensions as crew and tech support worked to achieve superhuman goals.

A postscript to the story is that the universe and its omnipresent intelligence may well have tried to warn NASA, through synchronicities, of this dangerous scenario *before* the flight (and during the first few minutes after lift-off). Before the launch, for example, a helium tank showed a higher-than-normal pressure. A member of the main crew, Ken Mattingly, was exposed to the measles and was scrubbed from the flight (and was replaced by Jack Swigert). A liquid oxygen vent-valve refused to close, and instructions had to be sent to the rocket several times before it could finally be shut. During liftoff itself, the center engine of the Saturn booster cut off two minutes early, requiring that the other four engines burn an additional thirty-four seconds.

Were all of these incidents part of a pattern of meaningful coincidences that was ignored? I will not endeavor to answer that in the face of so many extraordinary accomplishments by the thousands of team members at NASA. In light of what researchers know about intelligence—that it permeates reality—I have my own opinions about such matters. Matters of synchronicity call for the use of Measure and Assessment, and that, as we have already seen, is a matter of relativity.

What I *will* say is that the number thirteen is by no means an unlucky number. Some people might say that the misfortunes of Apollo 13 prove that the number is bad luck. My response is simple: "They made it back home against the odds."

By using all of our thirteen intelligences, *we make our own luck*!

Summary

The intelligences described earlier exist as absolute, definite units, but they can interact synergistically as seen in the examples of Terry Anderson and the flight crew of Apollo 13. The intelligences are always present in every action we perform, but they are not necessarily balanced (or focused upon evenly). Most people have an abundance in one or several of the thirteen intelligences. We use these particular abundances as strengths in order to survive and thrive.

Chapter Sixteen

Systems Thinkers Through The Ages

The following are some classical and obvious examples of realizing one's potential. Let us now focus solely on historical figures known to everyone, people who were able to realize potential in their lives to a degree that leaves most in awe. Doing so will help us to see that the theories we have discussed, and the NATI system in particular, exist in real life and not just as abstract concepts. Keep in mind that the goal of this book is to show people that certain dynamics exist in nature and are used by everyone. Let us therefore turn our attention to individuals who were able to activate potential energy in the living void in order to enrich mankind. As you will note, these individuals were/are capable of thinking abstractly, but in Whole Concept form.

Thomas Edison

I want to begin with Thomas Edison since his life provides an excellent model for the ideas we've been discussing. Born in 1847, Edison was the son of normal parents who had no special technical or scientific background. His mother was a schoolteacher, and his father worked at several jobs, such as running a food store and selling real estate.

Edison was a curious child who, in the estimation of his early teachers, asked too many questions. His grades were not very good and he especially hated math. His mother placed him in two other schools, but Edison still didn't do very well and his grades were poor. Accordingly, most of his education was the result of home schooling by his mother. Unlike Edison's formal teachers, his mother encouraged his curiosity so that her son became a voracious reader, consuming the classics as well as dozens of books on science. By the age of ten, he had built his first laboratory in the basement of his family home.

Edison had poor hearing for most of his life, and his ears apparently sustained considerable damage when he was fifteen as he ran to catch a departing train. The conductor grabbed him by the ears in order to pull him up to the already-moving platform between cars. Edison said later that he felt something snap inside his head as the conductor pulled on his ears.

As he got older, Edison scavenged bookstores more and more frequently to find books on science and chemistry. He also collected chemicals and "junk" to be used in his experiments, filling his rented room with materials even before he had any notion as to how he would use them.

From these humble beginnings, Edison would go on to invent the telephone, the phonograph, the incandescent light bulb, the movie projector, the voting machine, the stock market ticker, and hundreds more. Not bad for a child that formal education labeled as addled and confused!

Interestingly, Edison resisted any attempts during his life to have an operation to repair his hearing, for he claimed that his partial deafness enabled him to concentrate better. (A correlation with Beethoven is obvious here, since the famous composer wrote some of his most notable work after becoming totally deaf.) One might wonder if Edison would have been nearly as productive if he had not been able to focus his attention so completely on the thousands of tasks he undertook. Could he have tapped into the vast ocean of potential—that living nothingness—if he had had better hearing? One thing is for certain, however: Thomas Edison directly confronted the concept of wu as he read more and more about science in his formative years. A passion was ignited that made him goal-oriented and enabled him to translate ideas into reality. From a state of non-being came the many conveniences that we all live with and take for granted in modern times.

Another factor that seemed to enhance Edison's creativity was his habit of taking brief naps during the day. Wearing his trademark white suit, the Wizard of Menlo Park would periodically lie down on a workbench for short "power naps," waking refreshed with new energy and new ideas. One of the most remarkable aspects of Edison's life is that he merely stumbled upon many of his inventions by accident, indicating that his unorthodox habits (frequent naps would hardly be tolerated in the modern American workplace) made him an intuitive thinker. New applications for old materials simply materialized in his thinking.

By far, the most striking aspect of Edison's great achievements is that he and his fellow workers created the world's first invention factory, which was no doubt an outgrowth of collecting materials and performing experiments in his basement and rented rooms early in his life. In setting up his workplace as an area devoted to the design and manufacture of inventions, it is clear that Edison made the realization of potential a priority in his life. He consciously set about to gather raw materials and create larger systems from them, always using mind power and reflection to better understand how to apply his considerable knowledge. He was able to constantly format raw bits of data and transform them into useful information. In turn, he could then use his inventions for practical application, such as the mass consumption of light, electricity, or communication. In true NATI fashion, Edison was able to function on several levels at once, with various components of his overall intelligence always in contact with each other. His mind was truly a holographic mind because he used all thirteen intelligences at the same time when he focused on something.

From this brief description of his life, we can see how Focus, Parts, Synthesis, Intuition, Mental power, Models, and Procedures were especially important, though he unquestionably used all thirteen intelligences in his work.

Leonardo Da Vinci

One of the greatest geniuses mankind has ever known was Leonardo da Vinci, born in 1452 in Vinci, Italy. He was the illegitimate son of a twenty-five year old peasant woman, Caterina. His father took custody of Leonardo and moved the child to his home, where his son had access to numerous scholarly texts. Showing considerable talent for painting, Leonardo was apprenticed to Andrea del Verrochio in Florence. Verrochio claimed that da Vinci was much better than anyone in his studio, including himself, and resolved never to paint again.

Beginning in 1482, Leonardo spent seventeen years in Milan in the service of the Duke, Ludovico Sforza. During these years, Leonardo engaged in both artistic and scientific pursuits. He not only painted, but he designed weapons and machinery for the Duke, including a prototype of the military tank. Other studies included geometry, anatomy, chemistry, architecture, and flight. His sketches for a flying machine are now legendary since his study of flight mechanics was centuries ahead of its time. More than any other figure of his time, Leonardo da Vinci epitomized the Renaissance Man. Throughout his career as artist, scientist and inventor, he quite literally explored thousands of ideas and concepts. One idea would beget another; one path of inquiry always caused him to start new investigations in a never-ending cascade of information, understanding, and application. He, too, was an intuitive thinker, with several different aspects of his intelligence functioning simultaneously. He frequently locked himself in his room or workshop in order to simply think, to engage in speculation for its own sake. Here, too, we see a mind discerning the necessity for focus and concentration. It was almost as if Leonardo were engaging in a self-imposed deafness so that he could have greater clarity of thought. In essence, he voluntarily entered the living void to draw upon potential.

Like Edison many years later, we see in Leonardo a boy who was given access to abundant reading materials, and while we may speculate that both Edison and da Vinci were genetically predisposed to invention and genius, it is also equally clear that, early in their lives, both men set goals to accomplish as much as possible. Additionally, Leonardo's studio in Milan, always busy with apprentices laboring on multiple projects, might well be termed an invention factory in its own right. This kind of organization once again illustrates a mind passionate for potential, an intelligence functioning across a continuum of activity, with further potential as much a goal as the actual artwork or inventions. In Leonardo, we see a holographic model of intelligence. In modern terminology, Leonardo was one of the original "multitaskers."

We see in Leonardo an especially strong example of Beliefs and Concepts as he took abstract ideas and turned them into prototypes for many things not "formally" invented or used until the twentieth century.

The Wright Brothers

Orville and Wilbur Wright were the sons of Bishop Milton Wright and Susan Catherine Wright. Wilbur was born in 1867, Orville in 1871. Their parents created an atmosphere conducive to learning in their home. Of his childhood, Orville wrote, "We were lucky enough to grow up in an environment where there was always much encouragement to children to pursue intellectual interests; to investigate whatever curiosity aroused."[1] For a third time, we see information gathering encouraged by parents for its own sake, creating a pool of knowledge while simultaneously fostering the goal of realizing potential.

The life of the Wright brothers was not free from adversity despite a well structured, loving home life. In 1886, Wilbur was hit in the face with a wooden bat while skating on a frozen lake. Shortly thereafter, he became stricken with heart palpitations. Instead of attending Yale as his parents had intended, Wilbur remained at home for the next four years taking care of his mother who had tuberculosis. Wilbur remained an avid reader, however. By 1878, he and his brother had already been given a unique toy, a small flying machine made of bamboo and cork. This ignited the spark that led to flight. By the 1890s, Wilbur had read every book and paper he could track down on flying. It was during this period that he became obsessed with human-powered flight. He was in direct contact with *wu* and began directing most of his energies toward making flight a reality, or *yu*. Even before their first flight at Kitty Hawk, Orville and Wilbur made a kite out of wood, wire, and cloth. Based on this model, Wilbur was more certain than ever that an aircraft capable of lifting man off the ground was a real possibility, and it is here that we see a mind totally engaged with potential, allowing it to direct his actions in order to produce something from nothing—the airplane, a machine that had never existed before in the history of civilization (except, of course, in the mind of Leonardo da Vinci!).

The brothers built their flying machine while operating the Wright Cycle Company. In essence, this was their invention factory. The mechanical principles involved in the simple motion of a bicycle were coupled with intuitive thinking to produce the airplane. While they often experienced apparent dead-ends in their thinking, as well as small catastrophes in their field tests, these setbacks became opportunities for problem solving. The brothers were simply not going to be deterred. Their single-mindedness and devotion to flight—their focus if you will—was too strong to admit permanent failure. When published data on aerodynamics proved unreliable, they actually built their own wind tunnel to test possible architectural designs for their plane.

Although their initial flights lasted only a few seconds, the Wright Brothers would eventually succeed in longer flights that gained the attention of the United States Army. The rest, as they say, is history. In less than a century, man would enter the jet age and land on the moon, all thanks to the abiding passions of Orville and Wilbur Wright.

Along with the other intelligences, we see the Wright Brothers placing a premium on Models as they tried hundreds of designs that led to the first airplane.

Wolfgang Amadeus Mozart

Some regard Mozart, born in 1756, as the greatest musical genius of all time. His father Leopold was a fine musician in his own right and noticed with keen interest that his three-year-old son was able to pick out melodies on the piano. Later, at the age of six, Mozart was already composing, and his first symphonies were completed before his adolescence. Deemed a piano virtuoso by his father, Mozart was deprived of a normal childhood owing to the fact that he traveled extensively to entertain at the courts of Europe.

Undeniably, Mozart had the encouragement and tutelage of his father, and it is probably safe to say that Mozart was born with his talent already at a high stage of development. In this sense, he was extraordinary. While he no doubt benefited from the complex harmonies evolving in eighteenth century music (he patterned some compositions after Joseph Hayden, proving that even genius does not spurn sound models), it is obvious that Mozart was in touch with the void of all potential at an unusually early age.

Perhaps a clue to his genius (and one which is of no small importance in the NATI paradigm) is that he seems, in the estimation of many music scholars, to have had a natural precision and lyricism in his music, as if he were modeling not just Hayden, but nature itself. It is this latter trait that is most intriguing about the composer. His Thinking & Intelligence were *natural.* They were both existent *and* evolving almost from the time of his birth. It was as if he were born with a cosmic switch in the "on" position, making it possible for him to have easy intercourse with the realm of all possibilities.

Was there anything in his life comparable to an invention factory, as was the case with Edison, da Vinci, or the Wright brothers? Not in the conventional sense, but a strong case can surely be made that Mozart's own brain was an invention factory, with both hemispheres fluently communicating with each other, enabling him to produce harmonics and melodies that stunned the ears of his listeners. The instruments that he used were not new, of course, but the way he employed the classical orchestra was nothing short of revolutionary. In assembling raw data and applying it to large, complex models, such as symphonies and operas, Mozart synthesized and applied his accumulated knowledge in a natural, facile way.

In holistic terms, he was operating at peak efficiency for virtually his entire life. He was master of multiple forms: the symphony, string quartet, piano sonata, aria, concerto, and many others. In all of his work, regardless of genre, he was able to understand intuitively the complex mathematical relationships that underlie classical composition. To grasp all of these factors simultaneously—melody, instrument, genre, harmonics—is certainly the quintessential example of "whole mind" thinking.

In Mozart, therefore, we see that the application of intelligence is most fruitful when it follows a natural model. Potential is achievable when its pursuit unfolds according to fundamental principles spun into the very fabric of nature.

We can also examine the lives of many contemporary people to see dynamic levels of creativity and thinking.

Norbert Wiener

In 1948 Norbert Wiener, a professor of mathematics at the Massachusetts Institute of Technology (MIT), published a profoundly influential book called *Cybernetics* (MIT Press, 1948). The word 'cybernetics' comes from the Greek word *kybernetes*, meaning a steersman, a pilot who steers a boat. The word 'governor' comes from the same root. Wiener defined cybernetics as "the science of communication and control in animal and machine."[2] It has made a major contribution to the study of systems. Cybernetics focuses on how a system functions, regardless of what the system is—living, mechanical, or social. It tantalizes us with an ambitious promise—to unite different disciplines by showing the same basic principles are at work in all of them. Wiener proposed that the same general principles that controlled a thermostat can also be seen in economic systems, market regulation, and political decision-making systems. Self-regulation of systems by feedback, defined by Wiener as "a method of controlling a system by reinserting into it the results of past performance,"[3] became an engineering principle and was taken up by nearly all aspects of technology. If you can fully control one variable in a process, you can indirectly control them all by building in feedback links.

William Harvey

Turning to medicine and physiology, perhaps the next major step forward was by the English physician William Harvey who discovered the circulation of the blood. He published his ideas in 1628, showing how the heart pumped the blood around the body, finally refuting the prevailing theory, dating back to Galen in 170, that the liver was the central organ of the circulatory system and moved the blood to the edge of the body to form flesh. The heart and blood vessels are indeed a system, and they make a circle. The continuous change and movement of the heart and blood from moment to moment keep our internal environment steady.

Medicine has slowly unraveled many of our bodily systems since then, not only showing how each is a homeostatic system—that is, it regulates itself—but also how the various systems all fit together for the whole to work. With the new science of psychoneuroimmunology that has developed since the mid-1970s, we are discovering how body and mind work together as a larger system, how stress and emotional trauma can leave us susceptible to disease, how thoughts affect us physiologically in the form of neurotransmitters, and how actions of drugs are dependent on our belief in them, as shown by the placebo effect.

James Watt

The next installment in the story of systems came with James Watt. Born in 1736, he increased the power of the existing steam engines in. two ways. First, he designed a separate condensing chamber, which prevented loss of steam in the cylinder. Secondly, and more importantly from our point of view, in 1788 he invented the centrifugal governor, a revolutionary feedback device that automatically regulated the speed of the engine, for by adjusting the governor, the driver could control the engine to a steady rate.

The governor consisted of two lead balls that swung from a small pole. The influx of steam caused the pole to rotate and the balls to spin. The faster the spin, the higher and faster the balls rotated, like a fairground carousel. They were linked to a valve on the pole that adjusted the amount of steam and so controlled the speed of the spinning pole. The higher the balls spun, the more they closed the valve and reduced the speed. This allowed a much greater degree of control than had been possible in the past.

Jay Forrester

In 1961, Jay Forrester applied the cybernetic principles to the problems of economic systems, urban industry, and housing in his influential book *Industrial Dynamics* (Productivity Press, 1961).[4] Forrester's work later broadened to include the study of other social and economic systems using computer simulation techniques and is known as the field of system dynamics.

Forrester used a computer model in his book *Urban Dynamics* (Productivity Press, 1969) to try to understand the causes of urban growth and decay.[5] If it is possible to model the complex interplay of forces in cities, then perhaps the principles could be extended even further. It was this possibility that led the Club of Rome to sponsor a conference in 1970 on "The Predicament of Mankind." Forrester and his colleagues began to design a world system dynamics model that was rather like a global spreadsheet. The result was the hugely influential book

The Limits of Growth by Donella Meadows et. al.[6] The book looked at possible relationships between pollution, population, and economic growth, and used a computer model to draw conclusions about a sustainable future, conclusions that were (and still are) highly controversial. Their basic thesis is that if present growth in world population, pollution, food production, and use of natural resources goes unchecked, the limits to growth of the Earth will be reached within one hundred years. Its sequel, *Beyond the Limits* (Earthscan Publications, 1992) took the analysis further and offered slightly more hopeful conclusions than the original text.[7]

And Everyone Else!

What did all of these individuals have in common? What made them all similar despite their apparent uniqueness?

It is obvious that they were all able to focus keenly on their various endeavors. Furthermore, they were able to accumulate knowledge and build models based on established designs before moving on to more original concepts. They also endured a certain amount of hardship, transforming weaknesses into potential strengths. They followed constructive patterns of work, but were always able to transcend simple cause and effect in order to jump to the next level of functioning and operate intuitively so that spiritual energy could manifest itself. From an early age, they all possessed keen passions relative to their interests and had strong desires to accomplish, grow, and learn. Most of all, they operated holistically, using several different natural intelligences simultaneously. In doing this, each of these geniuses reflected the natural intelligence of the universe. Indeed, it would not be inaccurate to say that each and every creative thinker who has ever existed has cooperated with nature's grand design to foster human evolution.

It is also more than mere coincidence that each of these remarkable figures exhibited Maslow's eight factors critical to development. All had great intelligence and a strong sense of identity. Their development, which included virtue and a strong sense of character, was an ongoing process. Finally, each moved past doubt and adversity to relish peak experiences of joy, fulfillment, or transcendence.

Does this mean that only a select few are destined to become self-actualized? Does this imply that only people who are genetically predisposed to the attainment of great feats—or people with backgrounds that encourage learning for its own sake—can utilize the dynamics of potential? Absolutely not! The revolutionary vision of Natural Thinking & Intelligence is that every human being on the face of the earth has the ability to tap into the vast ocean that the Tao calls potential. Just imagine a world where more and more people begin to activate their dormant abilities. Such an image is dizzying, and when that day arrives, we will truly be heading toward the Omega Point.

SUMMARY

Geniuses have been recognized throughout history. In this chapter, we looked at the lives of Thomas Edison, Leonardo da Vinci, the Wright brothers, Wolfgang Amadeus Mozart, and many others. These were all people who realized a high degree of achievement because of the way in which they used their thirteen intelligences. While the NATI system does not promise to make everyone famous, it does show that the very same principles that make anyone a genius are present in every human being.

Chapter Seventeen

Achievement Models And Case Studies

We have seen that polarities are everywhere. They exist within Derald Langham's geometric model of the cell, and they exist in every human being as well. They are part of the cosmos, and if we are to decode our potential, we must come to grips with them.

Earlier, we established the importance of polarities and saw how they operate in various scenarios. Let's look at another real life case from another area of life before exploring how all of this information comes together in the dynamic open systems that are our individual matrices.

A Giant Gap

By 1992, new Head Coach Ray Handley had taken over the New York Giants from the very successful Bill Parcels, who had guided the team to a Superbowl victory. Handley's inability to motivate the team was implied by a headline in the *New York Daily News:* GUT INSTINCTS ARE CAUSING ULCERS. Handley was a highly intelligent man with keen Mental intelligence. In 1992, he attempted to teach his defense, headed by all-stars Lawrence Taylor and Carl Banks, what amounted to a very cerebral defensive system. As the newspaper headline indicated, this intellectual approach did not go over well with the players. The crux of the problem was that the players were all very much Intuitively-oriented (such as linebacker Pepper Johnson), whereas Handley functioned Mentally and by-the-book. Not surprisingly, the Giants suffered a losing season that year. This is a classic case of how Mental and Intuitive functioning can conflict with each other. The results of Coach Handley's system were decidedly negative. While it may have had many redeeming features, Handley made no attempt to learn from the almost palpable team unrest. Weakness could not be turned into strength because there was no examination of the tension between polarities.

Handley was fired and replaced by the former head coach of the Denver Broncos, Dan Reeves. As the newspaper correctly noted at the time, good coaching is not possible without communication skills, and Reeves received high marks on this count because he knew how to connect with his players. Reeves taught them enhanced Physical intelligence, but he did it by helping players work on their fundamental skills, which was something to which they could easily relate. This truly represented a middle way. In fact, his approach was labeled "selectively aggressive." Far from being a contradiction in terms, this phrase demonstrated that Reeves was addressing the issue of polarity.*

He employed mental tutoring as part of his overall system, but he did so in a way that increased the Physical and Intuitive capabilities of his players. With Reeves, it was not a case of either/or.

Competitive Pressure

The next case history deals with a young female athlete—an ice skater—who was trying to deal with the tremendous pressure of athletic competition. The issue at hand was "How do I overcome pressure when I compete?" In response, my associates and I tried our assessment approach to decoding potential. We decided that it was very much a Creative (Focus, Concept, and Expression) aspect that was having an effect on her performance. Our overall game plan was to zero in on such issues as her Awareness and Motivation concerning each of the four Functional factors, as well as her Beliefs/Confidence concerning each of the four.

*Selection and aggression in football are typically different schools of strategy.

Finally, we asked her to rate her Expression of the four—that is, the way in which she acted out all four Functional aspects. In addition to asking her to rate herself, we also asked her two coaches and one parent to do the same independent of each other. After intense questioning and the finalization of a NATI Matrix, our conclusion was that she was rushing her actions. Consequently, at times there was no smoothness, grace, or balance in her performance. Rather, she was overwhelmed on the ice.

There were three fundamental reasons behind her rushing—one was Physical, one was Mental, one was Emotional. The Physical reason was that she had sustained an ankle injury some three years earlier. It was rather severe, and although she had recovered, she had never truly *emotionally* overcome her fear of the incident reoccurring.

Another factor—the Mental—was that this adolescent had adopted a belief in perfection. She clearly viewed herself as a perfectionist and her mindset was ingrained within that Pattern. Psychologically, when perfectionists are put into a situation where they cannot be perfect, the results are identical to what occurred with this young athlete. They become tremendously pressured, more so than most because deep down they know that it is literally impossible to perform perfectly. We had a hard time trying to explain this to the youngster, so eventually we created a new Pattern. What we did was to change her interpretation of perfection. We did this by using a concept that we called "Perfection through Imperfection." Very briefly, this can be explained by the following example.

If an airplane takes off from New York headed for Los Angeles, it has an ultimate objective of safely arriving in L.A. within a given time frame. In order to do this, the airplane and its crew have to follow a pre-determined path. The path, however, is never absolute. That is to say that a plane in flight "drifts." If it strays off its course by several degrees, the gyroscope kicks in and corrects the course, maintaining a straight and level course until it drifts to another side by a few degrees. Eventually, however, it will safely arrive at its destination. When that occurs, we say that, assuming all other things are equal, this was the perfect conclusion to a flight.

It is this kind of pattern that we gave to the young athlete. If one part of a routine was not up to par, she could compensate for the problem by getting "back on course" later in the routine. Even if she was not satisfied with an entire routine or practice session, she could make adjustments the next time she was on the ice. The new pattern literally provided her with a middle way between the polarities of perfection and imperfection. She took the emotional energy inherent in her fear and stress and used it to find a new way of approaching her goal of excellence in skating. We were able to overcome the Mental mechanics of perfection by introducing Emotional power. It took some time, but it eventually worked.

I should point out that the entire session with this young lady took less than eight hours to accomplish. Efficiency is another hallmark of the Decoding Potential Program.

The Laws of Elizabeth Smith

A number of years ago we conducted a seminar on decoding potential whereby individuals were asked to state the issues on which they wished to focus. At that time we were applying a very simple format to the program. We simply listed the thirteen intelligences in their respective groupings of Creative, Organizational, and Functional.

CREATIVE	Awareness
	Belief
	Expression
FUNCTIONAL	Physical
	Emotional
	Mental
	Intuitive

ORGANIZATIONAL Models / Laws
Mirror / Feedback
Parts / Details
Order / Process
Measure / Priority
Whole / Synthesis

We then put the stated issue up on a chalkboard and proceeded to relate the focus of the issue to each of the intelligences to see what we could glean from the exercise. At one point, an entrepreneur, Elizabeth Smith, raised a business issue. Liz's issue dealt with a pending partnership arrangement with somebody she had been doing business with to some degree for the previous six months. We addressed her concern, which involved doubts she had relative to forming the partnership. She could not be more specific, but there were some underlying issues that needed to be uncovered. We quickly went through the creative aspects of Awareness, Beliefs, and Expression, but with little discovery.

We then examined the Organizational Group, and the first issue of Laws really set off the bell. Apparently, Liz's potential partner did not want to formalize the relationship with anything more than an informal memorandum. While Liz liked this person, she really didn't know if he was ready for a partnership. Given the involved nature of the projects they would be working on, we quickly came to the conclusion that there needed to be some sort of formalized legal agreement put into place. Elizabeth assured us that this was going to occur.

Over a year later, I met her at another conference and she explained to me that she and her partner had finally signed an agreement, although a major problem had arisen shortly thereafter. Because she had insisted on a strong agreement, things worked out in her favor despite some legal entanglements. She reiterated that if she had not pursued the course we uncovered at the NATI clinic, she might have ended up in considerable trouble.

This was a case in which someone simply needed to use her inner "lens" to focus on the intelligence of Laws. This alone was sufficient to prevent serious legal problems with her partner farther down the line. Paradoxically, she achieved her goal by avoiding the potential for trouble. Focus on Laws gave her the insight and courage to rid her inner matrix of self-deception regarding this man.

Let me emphasize that this was done with a very simple format using a simple chart of the thirteen intelligences.

Moving Toward Adulthood

In the mid-90s, Terry Anderson and I were doing some Achievement programs together throughout the tri-state area. In Rye, New York, we did a seminar that clearly demonstrated the elegant simplicity of the NATI Potential Program to me, Terry, and other participants. About halfway through the program, we were discussing the practical applications of the NATI intelligences. We used everyday examples and proceeded to show how NATI can shed light upon circumstances and foster enhanced understanding. At one point, Terry brought up a personal issue to further explore the efficacy and efficiency of NATI. His thirteen-year-old daughter was, like most teenagers, anxious to grow up. She wished to start wearing make-up and was giving her father and mother a very difficult time about this issue. Terry's wife, Maddie, is a beautiful and intelligent woman with strong, fundamental, mid-Eastern Christian values. She did not want her daughter walking around looking like Bozo the Clown, as she put it. There was therefore conflict between mother and daughter. Having gone through something similar when my two lovely girls were growing up, I sympathized with her. We attempted to come to some sort of insight and direction regarding this matter utilizing the various intelligences, but with not much success.

After about ten minutes, an elderly lady in the back of the room said, "The answer is Models." We all stopped, turned around, and looked at her. Terry asked her to expand on what she was saying. She said again, "The answer is Models. What I mean is what you need to do is get somebody closer to her age— perhaps eighteen or nineteen—who will show her how to put on the minimal amount of makeup that will still be effective in achieving both ends of the spectrum to satisfy mother and daughter." Using someone closer in age, the old woman explained, would provide someone that Terry's daughter could better relate to.

We finished the discussion with Terry and his wife by selecting one of my own daughters as her model. Although slightly older than the model we had envisioned, my daughter had a strong connection with Terry's daughter and was able to successfully bring about a resolution that served all parties.

This was a classic case in which Models was able to address polarities: make-up/no make-up and young/old. Terry's daughter wanted to look older by using an amount of make-up that would probably have been inappropriate. (It certainly would have been to Maddie!) Finding a model that was young (but not *too* young) was the key.

I might add that, secondary to Models, Process was also involved since my daughter could explain how to be subtler in applying make-up while still appearing to be more grown-up.

Lucy In Love

One of the most rewarding experiences I have had in assisting people through the Achievement Program came in the earlier years of its development, in the fall of 1985. It involved a thirty-six-year-old woman named Lucy. Lucy was a widow with two small children (four and six years old) who had lost her husband Joseph within the past twelve months. Joseph had been a young man when he was killed in a tragic automobile accident. Lucy was referred to me by a mutual acquaintance who was a participant in NATI classes I was teaching at Elizabeth Seton College in Yonkers, New York. Apparently, she had been deeply in love with Joseph, and at the time of his death, had been married for approximately ten years. By all accounts, it had been a fairy tale romance and marriage. She expressed the usual issues that are attached to this type of tragedy, namely grief, disillusionment, confusion, and loss. She could not understand why this happened to her and what she could do to help herself and her children.

Issues of this sort are never easily dealt with. They have a finality to them that is always sobering to those who are directly involved, and Lucy's case was just that. I spent an entire day with Lucy trying to answer her questions and generate a state of mind that would enable her to effect development for herself and her family. My greatest success in this area has always been in relating circumstances to absolutes such as virtue. By that, I mean that the circumstances involving people like Lucy and their spouses are actually representations of something on a deeper and more profound level. In Lucy's case, it was the virtue of love that she had experienced in her union with Joseph. I tried to explain to her that, in the cosmic scheme of things, Joseph represented the manifestation of the virtue of love in her life and it wasn't actually Joseph that she was truly experiencing. It was the virtue of love. That is, Joseph was a concrete image, while love was the abstract. What I was trying to promote was that Joseph was an archetype of something that was very beautiful and prominent in Lucy's life. I know this sounds quite idealistic and abstract, and it certainly did so to me at the time, but it was where the 13 principles had led me.

After many hours, Lucy appeared to be somewhat improved in her emotions, but I could tell that she wasn't buying what I was selling. I suggested to her that the best thing she could do was to focus her belief system on the virtue of love and not on Joseph. I was careful not to denigrate her relationship with her husband, but rather to look at love as an abstract power source to help her get through it and find greater understanding and direction. She assured me that she would try this approach as best she could since it was really the only thing that made logical sense, although it lacked emotional satisfaction at the time. She left, and I felt good about the fact that I was at least able to give her some different insight. I was hopeful that she could gain something by it.

I did not hear from her for a good nine months. One day, I was starting a lecture series at Seton College and, much to my surprise, I saw her name listed on the attendance schedule. This was late summer of 1986. I started the class by simply introducing myself and before I could get any further, Lucy raised her hand from the back of the class and asked if she could say something. I naturally acknowledged her. She stood up and began to express her situation in detail, including the loss, grief, and suffering she had experienced. She also related our earlier meeting and how we had spent a good deal of time attempting to come to grips with her grief. She told the group that she had had great disappointment and skepticism when we had concluded our original session simply because my words had not been what she expected or wanted to hear. She wanted her Joseph in the flesh! Nothing else had mattered.

Lucy went on to relate how she had searched for love in an abstract form and how things had started to happen in her life, such as meeting new people who were genuine, compassionate, understanding, and warm. She also stated how she began to see her life and the lives of her family in a different light. The notion of "falling in love with love" rather than with another human being was starting to take on a real life for her, so much so that she had begun to experience periods of elation. She concluded quite remarkably with the statement that the last six months had contained some of the most gratifying and compelling experiences she had ever encountered and that she was eternally grateful for my assistance and for the discovery of Natural Thinking & Intelligence.

For the next several months I couldn't help but think about the theory of Schrödinger's cat. Schrödinger's Cat Paradox refers to a cat being locked in a box with a device that may or may not release deadly cyanide gas depending on a single event: the radioactive discharge of an atom. As long as no one looks in the box, there is a fifty percent chance that the cat will be alive. Once the box is opened, only one possibility will have manifested itself. The cat is either dead or alive. Measurement is obtained by opening the box.

There was, and is, a question in quantum mechanics as to the measurement mechanisms of reality. One of those is known as linear measurement, which involves several possibilities of resolution. The other is known as relativity, which deals with a singular result. In the case of Schrödinger's cat, the measurement before the box was opened to see if the cat was alive or dead represented the linear measurement. There were two distinct possibilities. Once the box was opened, however, the wave function collapsed and the only possibility that existed was the reality at that specific time. The cat was either alive or it wasn't.

This whole dichotomy of quantum measurement and its duality was, and always has been, an issue. Thinking about what occurred with Lucy, as well as others I have dealt with concerning a focus on abstract virtue as a resolution, brought me to an understanding of a measurement process that reasonably addresses this quantum conundrum. I realized that the linear measurement was not an option since there was only one reality: Joseph was dead. The *relative* measure started to make sense when I realized that what happens when life concludes is merely a change of energy form, as in the case of the Second Law of Thermodynamics. There was a part of Joseph that took on the energy of the virtue of love. Lucy eventually related to that energy in a very strong way. *What brought Lucy to the point of resolution was her search for the abstract energy form of Joseph, not the physical, material representation.* She was able to connect with this energy and bring it into herself successfully.

We can see in this case study the central ideas of Holographic Theory. I was able to get Lucy to regard Joseph as a manifestation of the virtue of love, which exists in timeless form in the implicate order. The resolution of her grief revolved around realizing that the infinite potential that we call love exists in a pure, perfect state. Being "in love with love" meant that she was able to regard the experience of being in love on an abstract level. For

Lucy, love was no longer a single event but rather part of a much grander scheme that transcended a single person at a particular time.

Lucy is clear-cut proof that the notion of an abstract realm—an implicate order, if you will—can have real applications if we are willing to entertain a paradigm shift in our lives.

Peak Performance Under Pressure

Clearly, one of the most valid accounts of peak performance under pressure came from golfer Ray Floyd. Floyd, an accomplished professional, was being interviewed when he was asked about performing under tremendous pressure. Floyd's response was classic in relationship to NATI's Patterns for Potential Program. In essence, what he stated was that pressure is a perception. How you view your undertaking has a direct effect on the amount and type of pressure produced. Further, Floyd adopted a very interesting mental posture. What he said was that he simply wanted to be in a position to be able to win on any given day. He didn't put pressure on himself to win every time. He considered himself no less of a golfer if he didn't win a tournament. His belief system was structured so that he felt that he would be in a *position* to win within a given number of times—nothing more.

Pressure, then, is a self-induced phenomenon that is directly connected to Beliefs, Awareness, and Spirit. The Belief aspect is directly related to Confidence, while Awareness relates to Motivation and Focus. Spirit patterns can reduce pressure by focusing on Mission, Task, and Purpose rather than on Ego or Fear.

The Human Character Formula is very relevant here as always. Belief (or Confidence) is paramount. The ultimate Expression is changed by virtue of Awareness of a *possible* outcome in the context of a larger purpose—to play a good game of golf. With Ego negated, the Current Operating Procedure changes so that the inner matrix is healthier. While this process works, it is not easy to accomplish. However, with persistence it works.)

In systems terms, pressure represents a tremendous amount of energy, but the energy can be redistributed so that the brain receives new information regarding the definition of success. Physical performance on the golf course then changes, showing once again that the interchange between energy, information, and matter can bring a system into equilibrium. The elements of Fear, Ego, Ignorance, and Self-deception, however, will keep a system closed every single time.

TV Frenzy

In an April 11, 2004 newspaper article concerning what is wrong with TV shows (*Missing the Big Picture*), we see two NATI intelligences prominently highlighted.

Under a section titled “What’s Wrong,” the first intelligence was “no imagination” (i.e., no concepts). It stated that once a hit show appears, other networks simply copy them (e.g. *CSI* and *Law and Order*). But “copy” is simply an archetype of “modeling.” That is why this process is successful, in most cases. Others have implemented it and it worked!

The USS Cole Attack

The same day, another article appeared, called “Attack on *Cole*: Missed Sept 11 Clues.” The article presented a classical case for the lack of (and need for) synchronized holistic thinking. It is important that we cite a portion of the article here.

A reconstruction of events shows that the FBI and CIA failed to recognize their significance to act in concert (holistically) to intercept them (two hijackers) because of miscommunications and legal restrictions on the sharing of CIA intelligence info with investigators at the FBI. Problems developed even though FBI agents and CIA officers were assigned to each other’s units. The reconstruction of events shows the importance of the two men who figured centrally in examinations of the government's failure to prevent the September 11 attacks was misunderstood before the attacks because investigators thought that the two were associated with the Cole bombing and were not connected to a plot to strike targets within the US before 9/11.[1]

This clearly shows the negative result of both closed systems and non-holistic thinking. The two agencies were separate and distinct, which is fine since Parts is a NATI intelligence as well. The problem was and is that they were not connected, or part of the greater whole. This was supposedly rectified by the Department of Homeland Security. Moreover, they fell victim to restrictive, closed thinking systems by the lack of integration.

Not looking beyond the immediate focus, the *Cole* action fostered another negative result: 9/11. Had at least one of the organizations opened its culture, they would not be presently going through the time, energy, and humiliation of finger-pointing! The CIA's actions always seem to result from an incorrect use of NATI intelligences.

Olympic Pressure

Bill Mills, the 1964 Olympic gold medalist, had another perspective on pressure and performance. As a Sioux Indian, Mills was subjected to extensive discrimination for much of his life, which resulted in an indomitable attitude within the man. As he neared the finish line of his Olympic 10,000 meter race, he was in third place. His conscious thought was that he was going to be defeated, but he wasn't going to "lose." Third place in the Olympics is certainly no shame. However, when he accepted what seemed to be the inevitable fact of his defeat, he suddenly found new strength and sprinted dramatically into first place and won a gold medal. This perception—that one can be defeated but not be a loser—released large amounts of potential within Mills. In his matrix, information was converted into matter and energy. Just as with golfer Ray Floyd, Mills put himself in a position to achieve and then capitalized.

This concept is also demonstrated in the business world. Practically all functional people—the doers and the achievers—experience pressure from time to time since there are many factors that cause pressure. Pressure is categorically identified in NATI as an Expression. Expression, as we have stated earlier in the book, is the sum of Focus, Motivation, or Awareness, plus Beliefs or Perceptions. In this context, pressure is derived from the way people Believe, what they Believe in, their Interest, their Awareness, and their Motivation. It is also measurable in an Organizational sense, meaning that there are levels or degrees of pressure and peak performance. Further, pressure can manifest itself in any one of four ways (or combinations thereof), namely materially, emotionally, mentally, and intuitively. The same can be said to hold true for peak performance.

Summary

Case studies reveal that the Achievement Sciences and its components can enhance potential in virtually any area where we choose to apply our focus. In business, athletics, or affairs of the heart, we saw in this chapter how individuals were able to overcome problems, pressure, limitations, and grief.

Potential is available to anyone and is not limited to any particular goal or problem. Potential can be realized at any time and always results from the existence of the intelligences in the potential intelligence.

Chapter Eighteen

Applications and Programs

Let's see what all this looks like in real time. The key to all personal and societal existence is development as a life objective. As we develop, everything and everyone around us develops simply because restrictions and weaknesses take on a meaningful dimension.* Change begins within the individual. As the Bible says, "The Kingdom of God is within you" (Luke: 17:21). In this respect, everything we then undertake is a teaching/learning experience. This concept leads us to a perfect path and a higher model of reality.

There are several key technological systems at work in NATI. It should be kept in mind that the various aspects of NATI are not only philosophical and scientific, but also technological. We do not need to know each of these areas or to practice them. They are automatically in effect. All we need is to be aware of what they are and how they impact our lives.

The concept of Development impacts individuals as well as societies, groups, whole nations, and indeed, the entire planet. Some of the specific systems that are at work technologically in our individual and collective lives according to NATI are summarized as follows:

1. **The A + B = C System:** Awareness + Belief = Character
2. **The Four Functional Systems:** Achieve things Physically, Mentally, Emotionally, or Intuitively.

*Here again, we note how polarity, restriction, and weakness, which are viewed by us as unrealized potential, are utilized to gain strength and achievement. We are again turning negatives into positives.

3. **The Mirror:** Whatever bothers us is actually a reflection of our deficiency.
4. **Polar System:** Everything has an equal and opposite. At the same time, a higher unifying force always exists.
5. **Organizational System:** All organization possesses six intelligences: Principles, Laws, Processes, Synthesis, Details, Measure, and Feedback.
6. **Great Restrictors:** Dysfunctions of achievement relate to four factors: Fear, Ego, Ignorance, and Self-deception.
7. **The Integration System:** This is part of both the systems and NATI philosophy. As a systems component, it represents reviewing issues from a viewpoint of all thirteen intelligence principles and Polarity at the same time. The philosophy is simple: focus on development.

NATI Programs

The following are some actual programs utilized in NATI exercises. This matrix is composed of six organizational intelligences, Models (Laws), Processes (Order) Assessment (Measure), Mirror (Feedback), Details, and Integration. Further, it also composed of any number of "conceptual characteristics," all of which appear in a polarized state (e.g. in vs. out, off vs. on, etc.).

Since these six organizational principles are gleaned from principles of biology (cellular development) they provide us with a natural model of formation. By applying the conceptual characteristics to each, a comprehensive determination of organizational habits, profiles, and systems become recognizable.

1) Organizational Matrix

ORGANIZATIONAL	Models	Processes	Assessment	Mirror (Feedback)	Details	Integration
Open						
Closed						
Inner						
Outer						
Accept						
Reject						
Subjective						
Objective						
Need						
Want						
On						
Off						
Personal						
Impersonal						
Flexible						
Inflexible						
Power						
Control						
Uniqueness						
Recognition						
Self Interest						
Self-less						
Strong						
Weak						
Literal						
Figurative						

2) ABC Matrix

A + B = C	AWARENESS	BELIEF	COMMUNICATION
Open			
Closed			
Inner			
Outer			
Accept			
Reject			
Subjective			
Objective			
Need			
Want			
On			
Off			
Personal			
Impersonal			
Flexible			
Inflexible			
Power			
Control			
Uniqueness			
Recognition			
Self-interest			
Selfless			
Strong			
Weak			
Literal			
Figurative			

This matrix is composed of three creative intelligences (correlating with the Human Character Formula): Awareness, Belief, and Character (communication). Further, it is composed of any number and types of "conceptual characteristics," all of which appear in a polarized state (e.g. off vs. on; inner vs. outer, etc.). By applying the conceptual characteristics to each, a comprehensive determination of creative habits, profiles, and systems become recognizable.

Abstract and Concrete Thinking

This is also about abstract vs. concrete thinking. We have consistently mentioned how intelligence is recognizing data and incorporating it into a whole picture. This constitutes whole systems thinking. Let's see how we can enhance our ability with abstract thinking.

Follow this: "Our subject is awareness". This is our sample statement.

A concrete thinker responds: "Awareness of what?"

An abstract thinker responds: "Okay!"

Awareness of "what" implies "I need to attach a concrete idea of a concept in order to understand it," while "Okay!" implies "Awareness itself is the point of reference!" The concrete thinker is then limiting their COP by requiring a further definition. The abstract thinker engages the "whole picture".

Here's another one: "Let's discuss our beliefs".

A concrete thinker responds: "Which beliefs?"

An abstract thinker responds: "Right!"

"Which ones?" implies something other than Beliefs. "Right" implies Beliefs is our frame of reference. An important point here is that Awareness and Beliefs are symbols or archetypes. They have a general meaning, not a specific one. They are open, not closed.

Here are more exercises:

The subject is: "Making mistakes; being wrong."

> Closed thinking: "Who is he to say I'm wrong?"
>
> Open thinking: "I'll check out that comment as feedback."

The subject is: How to do something."

> Closed thinking: "This is the way I do it."
>
> Open thinking: "What are the available models?"

This subject is: "Change."

> Fixed: "Hey, that's the way it is and that's it."
>
> Variant: "There is a better way."

3) How Does the Focus Principle Integrate and Function?

The following are issues germane to implementing Focus.

- What rules might this focus violate in relation to mission, purpose, polar characteristics, and directional judgment?
- Does the focus dictate a rote process (vs. a NATI development process)?
- Does the focus adhere to priorities?
- Are said priorities based upon integration into an overall notion?
- What conflict/anger emanates as a result of the focus?
- How does the focus principle fit the various concepts? Are there any violations?

When one follows given principles, achievement/potential is realized.

Focus Example:

1. Am I taking one side or another? If yes, then you are engaging in opposition.
2. Is this issue a priority in achieving or developing?
3. Is this something Material, Mental, Emotional, or Spiritual?
4. What bothers me about the whole thing is this . . .
5. Am I looking at all six organizing principles and how I feel about each?

4) Weaknesses and Strengths

This is a simple exercise for developing one's potential. On the chart below, list your weaknesses and strengths.

MY WEAKNESSES	MY STRENGTHS
____________________	____________________
____________________	____________________
____________________	____________________
____________________	____________________
____________________	____________________

First, examine your weaknesses in terms of the Great Restrictors (Fear, Ego, Ignorance, Self-deception). Then, two parts follow. First, devise a workable plan of action for addressing and overcoming weaknesses. Then follow the plan as best as you can. Second, match your strengths to your weaknesses so that they can be utilized to overcome the weaknesses.

A secondary exercise is comparing both strengths and weaknesses to the Great Restrictors to determine if you are kidding yourself! *In other words, are you being honest about what your strengths and weaknesses really are? Perhaps having an outside party, such as your mate, boss, etc, confirm your opinion, might be an option.*

5) The Basic Matrix

Creative/Planning

- Awareness
- Beliefs
- Communication

Organizational

- Models
- Processes
- Assessment
- Mirror
- Details
- Whole

Functional

- Physical
- Mental
- Emotional
- Intuitive

This is a simple application model. Simply take an issue, define it as clearly as possible, and then investigate the issue from the standpoint of each of the intelligences above. Write down your thoughts about the issue from each of the perspectives. This should bring the underlying problem to the surface.* If not, ask

*This occurs by means of the fact that we are dealing with thirteen fundamental principles. This process can be likened to the spokes of a wheel with thirteen numbers. Sooner or later you will hit the winning number and you will know that you did. It is important that one implements the NATI language concepts in this process.

someone to go through it with you again. Keep track of each issue you undertake and eventually you will see definite paradigms occur . . . and finally a matrix embracing your nature!

The following is a case study utilizing the above format. It is quite generic and applicable for explanatory purposes.

Awareness	I'm pretty much aware of what's going on here and fairly well focused on it as well.
Beliefs	I believe in what I'm attempting to do, and I have confidence.
Expression/Communication	Things are not happening.
Models	I am imitating a well-conceived format, but maybe I'm making some modifications.
Process	The procedures being employed are standar
Assessment	I'm giving this issue a reasonable priority during my course of action.
Mirror	The feedback is: Well, I don't agree with what another party is proposing
Details	I think I have all the parts covered.
Whole	The whole picture is fine; however, everything is not coming together.
Materially	Seems to be okay.
Mentally	Except for a few things, we seem okay.
Emotionally	I'm a little concerned about . . . ?
Intuitive	Something is not exactly okay.

The issues to consider here are:

Models—The response from models above exposes some modification to our model (perhaps where things are going awry).

Mirror—The response from Mirror (above) exposes a potential problem with control. This latter principle is showing up within the Emotional and Intuitive intelligence.

6) Model for Optimizing Our Direction

The following are nine steps for achieving potential and setting our inner matrices in supreme order. Follow this format and you will expand your horizons.

- Be objective, flexible, and open to anything.
- Follow a proven plan.
- Take things impersonally.
- Be selfless and unconditional.
- Use the principle of development of potential as a life mission and focus.
- Synthesize and integrate information and events into an overall model.
- Treat weaknesses as potential. They need to be developed.
- Use Mirror/Feedback as insight to what is wrong or weak about your personal matrix.
- Stay focused and centered on life priorities, virtues, etc. as they relate to your development.

Optimizing our Direction
Here are examples:

1. Am I flexible and open toward this issue?
2. Is there a plan in place that has proven successful?
3. Am I being impersonal in my approach here?
4. Am I looking at everyone else's side of the issue honestly?
5. Is my ultimate objective here to develop and achieve?
6. Does everything come together, or are there still unresolved matters?

7. Am I aware of and dealing with my weakness in establishing my direction?
8. What bothers me most about this? Look inside myself for the answer.
9. I'm trying to find the voided virtues or the correct ones to employ.

7) Potential Intelligence Matrix – (PI)

Assess your Potential Intelligence of an issue using the following chart:

Concerning a Specific Issue	Column A	Column B
I am	Open	Closed
I am oriented	Externally	Internally
I am	Accepting	Rejecting
I am	Objective	Subjective
I take it	Impersonally	Personally
I am	Flexible	Inflexible
I feel	Strong	Weak
I feel	Positive	Negative
I view it	Figuratively	Literally
I view it	Abstractly	Concretely
My interest is in power.	No	Yes
My interest is in control.	No	Yes
My interest is in recognition.	No	Yes
My interest is in my uniqueness.	No	Yes
My interest is in being part of the group.	No	Yes
My interest is in casting an opinion.	No	Yes

The more answers you selected from column B, the better your potential for resolving the issues.

8) Format for Organizational Determination

- Is the Model centered on developing potential?
- Is its pattern open or closed?
- Is its assessment complementary?
- Is info being treated as feedback?
- Are Details (coherent units) probable or uncertain?
- Is it integrated with both positives and negatives?

CHAPTER NINETEEN

OF VALUES AND VIRTUE: THE ACHIEVEMENT VIRTUES

I have saved the best for last!

WHEN ALL ELSE FAILS YOU

We have spoken of how using the thirteen intelligences in a synergistic fashion is analogous to a symphony, where notes or instruments are blended to produce a synthesis of various parts. It is not an exaggeration to say there will never be a point at which composers will scratch their heads and say, "Oh well, it's all been done! There are no more symphonies to write." If possibilities are endless for composing a musical score with a finite number of notes (think of a piano keyboard—it is long, but there are only so many keys to work with), just imagine the possibilities that exist for the human brain to generate ideas using its trillions of cells. The history of mankind's evolution indicates that there may well be no limit to potential. If there is an endpoint to mankind's overall development, however, it may be similar to Teilhard de Chardin's Omega Point, where human consciousness will be a single entity—not a population of individuals. Even if this is the case, one may ask the question: What then? Will the collective consciousness of humankind be faced one day with new kinds of choices, such as the exploration of different dimensions? Indeed, will this new entity be in a position to *create* new dimensions or universes? Will the finished product of evolution, as many mystics and scholars have said for centuries, be part of what we now call "the Godhead"? These are tantalizing concepts depicting an abstract, transcendental, and yet scientific existence.

Actually, the good news is that we already have access to absolutes that can help us transcend the limits of our everyday lives to one degree or another. Yes, the NATI intelligences are absolutes, but what I am referring to is virtue, pure and simple. I am not talking about Sunday school mentality here, nor are you, the reader, in peril of receiving a sermon. As you know, the 13 principles are completely non-doctrinal, and I suggest that virtues are irreducible in nature and can, in and of themselves, help to unlock a person's potential.

Plato and You: Virtues over Values

We realize the best results in achieving our goals when we follow patterns of nature, patterns that naturally lead to development. Using concepts of potential development as the prime motivators for your very existence—the explicate manifestation that is you—can literally change your life. The same holds true for focusing on values and virtues and then adhering to them. As Socrates said, "Virtue is its own reward." This is why I firmly believe that when you attain virtue, you have already begun to realize potential in your life. The reasoning behind this is that virtue is a pure, clear path for pursuits.

I am, of course, making a fundamental distinction between values and virtues, for what one values in life is not necessarily a virtue. Simply consider people like Hitler or mass murderers like Ted Bundy. Values can be terribly skewed by the Great Restrictors we have alluded to so many times. The restrictors can corrupt our values and the matrix in which they exist. Values, however, tend to be less concrete than virtues, which is why adopting solid values is extremely important in one's struggle toward development of values that reflect virtues. In his book *Seven Habits of Highly Effective People*, author Stephen Covey says that it is essential to be "principle centered." It is extremely beneficial to find important principles and values that you can believe in and then bring them into your everyday life. In effect, they can become the "seat of your soul." In NATI terminology, this is the basis of your matrix, who you are and/or where you are going!

Virtue, as we said, is another matter since virtue is absolute, and is intrinsically related to Plato's concept of Ideal Forms. A virtue can never be reduced any farther than its given definition, nor corrupted, as values can be. Examples of virtue are abundance, acceptance, patience, tolerance, endurance, balance, consideration, clarity, courage, strength, fortitude, honor, honesty, truth, integrity, caring, cooperation, love, beauty, elegance, refinement, and confidence. I consider all of these to be the achievement points of excellence that define a superior path to achievement, change, and growth. Significant power is mobilized in one's matrix when these virtues are implemented. In chapter three, we chronicled many contemporary ills—problems in education, government, and religion—in showing the universal need for the Achievement Sciences. Consider for just one moment how various institutions—open systems, if you will—could be changed for the better with the power of virtue as a driving force behind man's creativity. Individually, our COPS would be changed because Core Human Dynamics would be used in a more positive manner. Power and control would be used to accept people and their ideas. Drive and motivation would be aimed at the common good. Attention-seeking would not lead to ego, but rather to a healthy self-image and a feeling of uniqueness. Value judgments would be fair and unbiased, with creative expressions stemming from a focus on (and belief in) virtue!

If you think all of this is too idealistic, you might be right. However, ideals are attributes that can direct our matrices away from restrictive notions and provide a clear direction. Consider some of the advantages of pursuing virtue:

- They give strength, are positively oriented, with no negative feedback, either consciously or unconsciously.

- Inefficient patterns are put into perspective, enabling one to discard distractions and focus on goals.
- Virtues flow naturally, reducing stress since focus on them distracts the mind from tension.
- They are self-sustaining. They provide confidence and motivation in public and personal life because they cannot be jaded or transcended. They are strengths in themselves and have no agenda.
- Virtues are self-organized. They innately lead to better intuition, organization, and functionality. They continually interconnect in a positive manner.
- They are always practical. They are easy to follow because they are identifiable. Following them becomes second nature.
- They produce expanded awareness and can see all parts of a situation. They rise above duality (that is, the either-or mentality/quantum measurement).

Some may believe this is all very impractical, but today pragmatism is very corruptible and is, in many cases, if not most, the easy way out.

Practicality of Virtue

At this point a practical example is in order. Not too long ago, I was taking a golf lesson from my friend, John Kennedy, head golf pro of Westchester Country Club in Rye, New York. John and I are always discussing ways golfers can achieve a greater degree of potential. I was in a terrible slump and couldn't break out. During the lesson, John asked me how this book was coming along and what were some of the keys. When I mentioned virtues and their application, something struck me. I was not applying the notion of virtue during this golf slump. I began to question John with the objective of reducing the mechanics of the golf swing down to what virtues may be at the root cause of my poor swing execution. It didn't take long to discover that the two primary *missing* virtues were patience and trust. For you golfers, I was not "waiting" on my swing, and I wasn't trusting it either. After one round of golf while focusing on these two virtues, I was almost totally back to my better self.

Business and professional applications are no different! I have come to see that failure, mediocrity, etc. are due to a lack of some virtue.

Virtues as Systems

We discussed systems at great length in earlier chapters. Virtue can enhance virtually any system—personal, societal, or cultural. Notice how many of the characteristics mentioned in the section above relate to General Systems Theory! Virtues, when incorporated into *any* system or matrix, behave like dynamic components. They flow naturally and are self-sustaining and self-organized. They are also balanced, the very heart and soul of complementarity. They constantly give strength, supplying energy, a vital component to any system. All of this is to say that virtues are not subject to any laws pertaining to dissipative structures, such as the Second Law of Thermodynamics. The reason for this is that they are absolute. They cannot be broken down or changed. They are incorruptible. This is why I emphasize their potential and transformative powers.

The question, therefore, is not so much whether or not we should adopt virtues. That is a given. Rather, what we should ask ourselves is "Can we afford to live without them?"

And Finally . . .

The word "natural" is prevalent in today's society. We want foods that are natural and free from harmful chemicals, preservatives, and pesticides. We want herbal remedies, aerosols free of CFCs, asbestos-free and lead-free building materials, natural exercises (such as yoga), natural diets, cleaner fossil fuels, solar energy . . . and the list goes on. In short, we want things that work in harmony with nature in order to be healthy and live longer lives. But aren't these goals? Of course they are. We have a natural tendency, conscious or not, that wants to realize potential. And this is the revelation of Achievement Sciences: there is a natural intelligence *within* us, a natural way to think—and there is no activity in the entire world that we can perform without thinking. If we crave things that are natural, then we need look no farther than our own thought processes.

The Thrill of Victory and The Lessons of Defeat

Although the decoding of potential has set forth scientific and philosophical means for touching deeper levels within ourselves, the development does not always come easily. Achieving one's potential is akin to the undertakings of an Olympic athlete. The pains of Olympic training are very much in league with what some of us need to go through in order to achieve desired results. Focus and Belief are not just words. They are calls to action, so to speak. Moreover, defeat and failure are also part of the game. We learn by experience, to some degree, but only if we accept the negative feedback of failure and prepare ourselves for its reoccurrences.

Commitment to the development of potential is far and away the most important responsibility of all humankind. Nature has given us all the natural abilities we need. However, this not the end of the story, for such a book as this can have no end without thinking on some level. You, the reader, are the logical conclusion to these chapters . . . and you have just begun.

Summary

While values can be skewed by the Great Restrictors, which can corrupt our inner matrices, virtue is absolute. With virtue, our COPs can be changed since Core Human Dynamics can be used in a more positive manner. Virtues give strength, put inefficient patterns into perspective, flow naturally, reduce stress, are self-sustaining and self-organized, are practical, and expand awareness. In short, they have transformative powers. Don't be afraid to use them!

Addendum
THE SCIENCE OF ACHIEVEMENT MATRIX

NATURAL INTELLIGENCE	*NATURAL THINKING*
INVARIANT PRINCIPLES Relates directly to seeing things as they are	**VARIANT PRINCIPLES** Relates directly to viewing things as one sees
Polarity 1. Positive (+) 2. Negative (–) 3. Neutral (0) Direction	**The Great Restrictors** (What Blocks Our Achievement) 1. Fear and Anger 2. Self-Deception 3. Ignorance/Confusion/Doubt 4. Ego
The Human Character Formula A Awareness [A] (Information) +B Beliefs [B] (Knowledge) =C Character [C] (Understanding)	**Core Human Dynamics** (Our Natural Motivation) 1. Power 2. Control 3. Acceptance/Inclusion 4. Uniqueness/Exclusiveness/One-Upmanship 5. Attention/Recognition-Seeking 6. Self-Interest 7. Judgment/Values/Appraising 8. Motivation/Drive/Will
The Four Human Functions 1. Physical 2. Mentally 3. Emotionally 4. Spiritually/Intuitively/Instinctively	**The Absolutes** – Polarity – The 13 Principles – Virtues – Potential
The Six Ways Humans Organize 1. Models, Laws, Rules 2. Procedures, Methods, Processes 3. Measures, Make Judgments and Appraisals, Priority 4. Give and Receive Feedback 5. Segregate, Divide into Parts 6. Integrate, Synchronize Combine	**Steps for Optimizing Comprehension, Clarity, and Reaching Our Potential** 1. Objectivity 2. Follow a Proven Plan 3. Be Impersonal 4. Unconditionality, Selflessness 5. Use the Principle of Development as Resolution and a Life Mission 6. Synthesize and Integrate Events and Information 7. Use Weakness as a Potential 8. Use the Mirror/Truth Seeking 9. Stay Focused on Values and Priorities

References

Introduction

[1] *Dictionary of Science & Technology*, New York: McGraw-Hill, 9th Edition, 2002.
[2] Ibid.

Chapter One

[1] Klaus Mainzer, *Thinking in Complexity* (Berlin, Germany: Springer-Verlag, 1994).
[2] H. R. Pagels, The Cosmic Code (New York: Simon & Schuster, 1982).
[3] Ibid.
[4] *Dictionary of Science & Technology*, New York: McGraw-Hill, 9th Edition, 2002.

Chapter Two

[1] *Howard Gardner, Multiple Intelligences: The Theory in Practice* (New York: Basic Books, 1993).
[2] *Oxford Companion to Philosophy* (London: Oxford University Press, 1995).
[3] *Webster's New World Dictionary*, 2nd College Edition, s.v. "Intelligence."

Chapter Three

[1] Marilyn Ferguson, *The Aquarian Conspiracy* (Los Angeles: J. P. Tarcher Inc., 1980).

Chapter Four

[1] Diane Dreher, *The Tao of Inner Peace* (New York: Harper Perennial, 1991).
[2] David Bohm, *Causality and Change in Modern Physics* (Philadelphia: Univ. of PA. Press, 1971).
[3] Ibid.
[4] Ibid.
[5] Ibid.
[6] P. Buckley and F. David Peat, *A Question of Physics: Conversations in Physics and Biology* (London: Routledge & Kegan Paul, 1979).
[7] Ibid.
[8] I. Bentov, *Stalking the Wild Pendulum* (New York: Dutton Publishers, 1977).
[9] W. Heisenberg, *Physics & Beyond* (New York: Harper & Row, 1971).
[10] Deepak Chopra, *Quantum Healing* (New York: Bantam Publishing., 1989).
[11] Bernie Siegal, MD, *Love, Medicine, and Miracles* (New York: Harper Collins, 1990).
[12] Deepak Chopra, *Quantum Healing*.

[13] Norman Vincent Peale, *The Power of Positive Thinking* (Englewood Cliffs, NJ: Prentice-Hall, Inc., 1952).
[14] Fritjof Capra, *The Tao of Physics* (New York: Bantam, 1976).

Chapter Five

[1] James Gleick, *Chaos* (New York: Penguin Books, 1987).
[2] Ann Weiser Cornell, *The Power of Focusing* (New York: MJF Books, 1996).
[3] Ibid.
[4] Wikipedia, the free encyclopedia
[5] Ibid.
[6] Ibid.

Chapter Six

[1] *Webster's New World Dictionary*, 2nd College Edition, s.v. "organization."
[2] David Bohm, quoted in *Looking Glass Universe*, John P. Briggs and F. David Peat (New York: Simon & Schuster, 1984).
[3] Ibid.
[4] M. Plank, *The Philosophy of Physics* (New York: Norton Press, 1982).
[5] Terry Anderson, *Den of Lions* (New York: Crown Publishing Group, Inc., 1991).
[6] Ibid.
[7] Karl Pribram, *The Holographic Principle* (Berkeley Heights, N.J.: Freeman Press, 1979).
[8] L. LeShan, The Mystic, *The Median and The Physicist* (New York: Viking, 1974).
[9] P. Buckley, and F. David Peat. *A Question of Physics: Conversations in Physics and Biology* (Toronto, Canada: University of Toronto Press, 1979).
[10] Brian Greene, *The Elegant Universe* (New York: Vintage, 2000).

Chapter Seven

[1] David Keirsey and Marilyn Bates, *Please Understand Me* (Del Mar, CA: Prometheus Nemesis Books, 1992).
[2] Rupert Sheldrake, *Seven Experiments That Could Change the World* (Los Angeles: Berkeley Publishing Group, 1995).
[3] Fritjof Capra, *The Tao of Physics* (New York: Bantam, 1976).
[4] Fred Alan Wolf, *The Dreaming Universe* (New York: Touchstone, 1995).

Chapter Eight

[1] Louis de Broglie, L. Armand, and P.H. Simon, *Einstein* (New York: Peebles Press, 1979).
[2] William Blake, quoted in *Major British Poets of the Romantic Period*, ed. by William Heath (New York: Macmillan, 1973).
[3] Diane Dreher, *The Tao of Inner Peace* (New York: Harper Perennial, 1991).

Chapter Nine

[1] J.S. O'Connor and Ian McDermott, *The Art of Systems Thinking* (San Francisco, CA: Thorson's Publishing, 1997).

Chapter Ten

[1] J. G. Miller, *Living System* (New York: McGraw-Hill, 1978).

[2] Peter F. Drucker, *The Essential Drucker* (New York: Harper Publications, 2003).

[3] Ibid.

[4] Teilhard de Chardin, *The Divine Milieu* (New York: Harper and Row, 1960).

Chapter 11

[1] Lawrence S. Bale and Gregory Bateson,, "Cybernetics and the Social/Behavioral Sciences," www.narberthpa.com/Bale/Isbale_dop/cybernet.html

[2] David S. Walonick, "General Systems Theory," www.survey-soft-ware-solutions.com/walonick/systems-theory.htm

[3] David Bohm from Science, Order, and Creativity, quoted in Keepin, "Lifework of David Bohm," www.vision.netau/~apaterson/science/david_bohm.htm.

[4] Will Keepin, "Lifework of David Bohm," www.vision.netau/~apaterson/science/david_bohm. htm.

[5] David Bohm, from *Science, Order, and Creativity*, quoted in Keepin, "Lifework of David Bohm.".

[6] Karl Pribram, *The Holographic Principle* (N.p.: Freeman Press, 1979).

[7] Will Keepin, "Lifework of David Bohm," www.vision.netau/~apaterson/science/david_bohm. htm.

[8] David Bohm, from *Science, Order, and Creativity*, quoted in Keepin, "Lifework of David Bohm."

[9] Ibid.

[10] John P. Briggs and F. David Peat, *Looking Glass Universe* (New York: Simon & Schuster, 1984).

[11] Albert Einstein, *Ideas and Opinions*, Sonja Bargmann, trans. (New York: Crown Publishers, 1954).

[12] J.D. Barrow, *Theories of Everything* (New York: Fawcett, 1991).

[13] Ibid.

Chapter Twelve

[1] David S. Walonick, "A Holographic View of Reality," www.survey-software-solutions.com/walonick/reality.htm.

[2] James Surowiecki, *The Wisdom of Crowds: Why The Many Are Smarter Than the Few and How Collective Wisdom Shapes Business, Economics, Societies, and Nation* (NewYork: Doubleday, 2004).

[3] Ibid.

[4] Ibid.

Chapter Thirteen

[1] *Dictionary of Science* (New York: McGraw-Hill, 9th Edition). http:/ /whatis.techtarget.com/definition/0 , , sid 9_ gci341236,00.html. Schrödinger's cat is a famous illustration of the principle in the quantum theory of superposition proposed by Erwin Schrödinger in 1935. Schrödinger's cat serves to demonstrate the apparent conflict between what quantum theory tells us is true about the nature and behavior of matter on the microscopic level and what we observe to be true about the nature and behavior of matter on the macroscopic level.

[2] John P. Briggs and F. David Peat, *Looking Glass Universe* (New York: Simon & Schuster, 1984).

[3] Karl H. Pribram, *The Holographic Principle* (N.p.: Freeman Press, 1979).

[4] P. Russell, *The Global Brain* (New York: St. Martins Press, 1983).

[5] David Bohm, *Quantum Theory* (Englewood Cliffs, NJ: Prentice Hall, 1951).

[6] James Gleick, *Chaos: Making a New Science* (New York: Penguin, 1988).

[7] Ilya Prigogine and I. Stengers, *Order Out of Chaos* (New York: Bantam Books, 1983).

[8] *Oxford Companion to Science* (Oxford, England: Oxford University Press, 1995).www-gap.des.stand.acuk/~history/Mathematicians/Einstein.htm. In 1905 Albert Einstein made important contributions to quantum theory, but he sought to extend the special theory of relativity to phenomena involving acceleration. The key appeared in 1907 with the principle of equivalence, in which gravitational acceleration was held to be indistinguishable from acceleration caused by mechanical forces. Gravitational mass was therefore identical with inertial mass.

[9] Karl H. Pribram, *The Holographic Principle* (N.p.: Freeman Press, 1979).

[10] David Bohm (1917–94) was one of the foremost theoretical physicists of his generation and one of the most influential theorists of the emerging paradigm through which the world is increasingly viewed. Bohm's challenge to the conventional understanding of quantum theory has led scientists to re-examine what it is they are doing and to question the nature of their theories and their scientific methodology. He brought together a radical view of physics, a deeply spiritual understanding, and a profound humanity. In the years before his death in 1992, Bohm lectured worldwide on the meaning of physics and consciousness. In an interview in 1989 at the Niels Bohr Institute in Copenhagen, where Bohm presented his views, Bohm spoke on his theory of wholeness and the implicate order. The conversation centered around a new worldview that is developing in part of the western world, one that places more focus on wholeness and process than analysis of separate parts. Bohm explained the basics of the theory of relativity and its more revolutionary offspring, quantum theory. Either theory, if carried out to its extreme, violates every concept on which we base our understanding of reality. Both challenge our notions. Bohm is not the only researcher who has found evidence that the universe is a hologram. Working independently in the field of brain research, Stanford neurophysiologist Karl Pribram has also become persuaded by the holographic nature of reality. Pribram was drawn to the holographic model by the puzzle of how and where memories are stored in the brain.

11 John P. Briggs and F. David Peat, Looking Glass Universe (New York: Simon &

Schuster, 1984). Drs. John Briggs and F. David Peat recreate the mind-boggling journeys taken by several prominent scientists. In lively, non-technical language, the authors describe how scientists in physics, chemistry, biology, and neurophysiology have hit upon theories that could revolutionize not only their disciplines, but the way all of us think about reality. These "looking glass" theories propose that we are, at this very moment, living in an Alice-in-Wonderland universe where each part is in fact the whole, where a scientist conducting an experiment is himself the experiment, and even inanimate objects contain consciousness. Finally, we learn how their theories may fit together into a single new hypothesis that gives scientific meaning to the ancient mystical idea that the universe is One.

[12] Harold Elliot Varmus, *Columbia Encyclopedia, 6th Edition* (New York: Columbia University Press, 2004). In a February, 2004 presentation to medical researchers at New York Medical College, Nobel Laureate Dr. Harold Varmus said a family of genes known as "ras" may be the single most significant component of runaway cancer cells. He said that maintenance genes tell the cells how to grow, what to attack, and what signals from other cells to ignore. "Cancer of the lung is the most common form of cancer among men and women in America, with about 70,000 deaths a year," Varmus said. "And at least 30 percent were found to have ras mutations. These genetic mutations are needed to maintain the integrity of the cancer cells, and if you interrupt their operation, you undermine the cell. If you take away the mutant gene, the cell dies." Oncogenes are normal genes that control growth in every living cell, but which under certain conditions can turn renegade and cancerous. Varmus and J. Michael Bishop's work, along with the work of a number of other research scientists, stem from the hypothesis that the growth of cancerous cells is not the result of an invasion from outside the cell, but rather a misuse of a normal gene by a retrovirus as a result of exposure to some aggravating carcinogen, such as radiation or smoke.

[13] Ludwig von Bertalanffy, International Society for The Systems Sciences. http://www.isss.org/lumLVB.htm

Chapter Fourteen

1 Institute for Personnel Development, *New York Times* article (Austin, Texas, Hoovers Custom News).

2 Ernst & Young, *New York Times* article (Austin, Texas, Hoovers Custom News).

Chapter Fifteen

1 Terry Anderson, *Den of Lions* (New York: Crown Publishing Group, Inc., 1991).

2 Jim Lovell and Jeffrey Kluger, *Lost Moon* (New York: Houghton Mifflin, 1994).

Chapter Sixteen

[1] Gary Bradshaw, "Wilbur and Orville Wright," www.warn.umd.edu/~stwright/WrBr/Wrights.html.

[2] Norbert Weiner, *Cybernetics*, (Cambridge, MA: MIT Press, 1948).

[3] Ibid.

[4] Jay Forrester, *Industrial Dynamics* (University Park, IL: Productivity Press, 1961).
[5] Jay Forrester, *Urban Dynamics* (University Park, IL: Productivity Press, 1969).
[6] Donella Meadows, *The Limits of Growth* (New York: Random House, 1972)
[7] Donella Meadows, *Beyond the Limits* (London: Earthscan Publications, 1992).

Chapter Nineteen

[1] Steven Covey, Seven *Habits of Highly Effective People* (New York: Simon & Schuster, 1990).

Bibliography

Anderson, *Terry. Den of Lions.* New York: Crown Publishing Group, Inc., 1993.

Aquinas, T. Summa. *Theologica in Selected Political Writings*, Ed. A.P. d'Entreves. Oxford:University Press: NewYork and London, 1948.

Ardrey, R. African Genesis. New York: Dell, 1961.

Asimov, Issac. *Science, Numbers & I.* Garden City, NewYork: Doubleday, 1968.

Bale, Lawrence S. and Gregory Bateson. *Cybernetics and the Social/ Behavioral Sciences.* www.narberthpa.com/Bale/lsbale_dop/cybernet.htm.

Banting, P.M. "Marketing, Scientific Progress, and Scientific Method." *Journal of Marketing* (1978),99-100.

Barrow, J.D. *Theories of Everything.* New York: Fawcett, 1991.

Bateson, Gregory. *Mind & Nature.* New York: Dutton, 1979.

Bentov, I. *Stalking the Wild Pendulum.* New York: Dutton, 1977.

Beukema, P.L. "Predicting Organizational Effectiveness with a Multivariate Model of Organic and Mechanistic Value Orientation." Diss. University of Southern California, 1974.

Bohlen, J.M., C.M. Coughenour, H.F. Lionberger, E.O. Moe, and E.M. Rogers. "Adopters of New Farm Ideas: Characteristics and Communication Behavior." *In Perspectives in Consumer Behavior*, eds. H.H. Kassarjian and T.S. Robertson. Glenview, IL: Scott Foresman, 1968.

Bohm, David. *Causality and Change in Modern Physics.* Philadelphia: University of Pennsylvania Press, 1971.

—— *The Enfolding-Unfolding Universe.* Interview by Renee Weber in Revision Summer/Fall, 1978.

—— *Quantum Theory.* Englewood Cliffs, NJ: Prentice Hall, 1951.

—— *The Physicist and the Mystic—Is a Dialogue Between Them Possible?* Interview by Renee Weber in Revision Spring, 1981.

—— *The Special Theory of Relativity.* New York: W.A. Benjamin, 1965.

—— *Wholeness and the Implicate Order.* New York: Prentice Hall, 1980.
—— and F. David Peat. *Science, Order, and Creativity.* London: Ark, 1987.

Boulding, K.E. *Evolutionary Economics.* Beverly Hills, CA: Sage Publications, 1981.

Bradshaw, Gary. Wilbur and Orville Wright. www.wam.umd.edu/~stwright/WrBr/Wrights.html.

Briggs, John P. and F. David Peat. *The Looking Glass Universe.* New York: Simon & Schuster, 1984.

Brown, R. "A Brief Account of Microscopical Observations." *Philosophical Magazine* 4 2000 161.

Buckley, P. and F. David Peat. *A Question of Physics: Conversations in Physics and Biology.* London: Routledge & Kegan Paul, 1979.

Burr, H.S. *Blueprint for Immortality: The Electric Patterns of Life.* London: Neville Spearman, 1972.

Cairns, Huntington. *Legal Philosophy from Plato to Hegel.* Baltimore & London: Johns Hopkins Press, 1949.

Calder, Nigel. *The Key to the Universe.* London: Penguin Books, 1981.

Capra, Fritjof. "*Bootstrap Theory of Particles.*" Revision Fall/Winter, 1981.

——. *The Tao of Physics.* New York: Bantam, 1976.

——. *The Turning Point: Science, Society and the Rising Culture.* New York: Simon & Schuster, 1976.

——. *The Tao of Physics*, 4th ed. Boston: Shambhala, 2000.

Castaneda, Carlos. *The Teachings of Don Juan.* New York: Ballantine Books, 1968.

Chopra, Deepak. *Quantum Healing.* New York: Bantam, 1989.

Cook, Theodore. *The Curves of Life.* New York: Dover, 1979.

Cooper, R.G. *The Performance Impact of Product Innovation.* European Journal of Marketing, 1984.

Cornell, Ann Weiser. *The Power of Focusing.* New York: MJF Books, 1996.

Costley, D.; K. Downey and M. Blumberg. "*Organizational Climate: the Effects of Human Relations Training.*" University Park, PA: Working Paper, Penn State University, 1973.

Covey, Stephen. *Seven Habits of Highly Effective People.* New York: Simon & Schuster, 1990.

de Broglie, L. and Armand, L, Simon. *Einstein.* New York: Peebles Press, 1979.

de Chardin, Teilhard. *The Divine Milieu.* New York: Harper and Row, 1960.

d'Entreves, A.P. Natural Law: *An Introduction to Legal Philosophy.* London: Hutchinson & Co., 1951.

Deshpande, Rohit, and A. Parasuraman. "Organizational Culture and Marketing Effectiveness." In *Scientific Method and Marketing.* P.F. Anderson and M.J. Ryan eds. Chicago: American Marketing Association, 1984.

Deshpande, Rohit, and F.E.Webster, Jr. "Organizational Culture and Marketing: Defining the Research Agenda." *Journal of Marketing* 1989 3 3-15.

Dinsdale, H. "Future Thinking." *Future Survey.* 1993 15 (4).

Donnelly, J.H. Jr. & J.M. Ivancevich. "A Methodology for Identifying Innovator Characteristics of New Brand Purchasers." *Journal of Marketing Research* 1974 11 331-334.

Dreher, Dianne. *The Tao of Inner Peace.* New York: Harper Perennial, 1991.

Drucker, Peter F. *Innovation and Entrepreneurship: Practice and Principles.* New York: Harper & Row Publishers, 1985.

____ *The Essential Drucker.* New York: Harper Publications, 2003.

Edwards, P. *Encyclopedia of Philosophy*, 1-8, In P. Edwards ed. New York: Macmillan Publishing Company and The Free Press, 1967.

Einstein, Albert. *Ideas and Opinions.* Sonja Bargmann, trans. New York: Crown Publishers, 1954.

—— *Out of my Later Years.* New York: Philosophical Library, 1950.

Etizone, A. *The Moral Dimension.* New York: The Macmillan Press, 1990.

Evan, W.M., & G. Black. "Innovation in Business Organizations: Some Factors Associated with Success or Failure or staff proposals." *Journal of Business* 1967 40 519-530.

Feldenkreis, S. *Potent Self.* New York: Harper & Row, 1985.

Ferguson, Marilyn. *The Aquarian Conspiracy.* Los Angeles: J.P. Tarcher Inc., 1980.

Fieldler, F.E. *A Theory of Leadership Effectiveness.* New York: McGraw Hill, 1967.

Flower, Robert J. *The Intelligence Cubes...and Their Patterns.* Aurora, CO: Gala Publishing, 1991.

Forrester, J. "Critical Theory of Planning Practice." J. Forrester, ed. *Critical Theory and Public Life*. Cambridge, MA: MIT. 1985 202-227.

Fuller, R.B. *Synergetics*. New York: Macmillan Press, 1975

Gardner, Howard. *Multiple Intelligences: The Theory in Practice*. New York: Basic Books, 1993.

Gleick, James. *Chaos*. New York: Penguin Books, 1987.

Globus, G, G. Maxwell. and I. Savodnik. *Consciousness and The Brain*. New York: Plenum, 1976.

Gold, B. "Technological Diffusion in Industry: Research Needs and Shortcomings." *Journal of Industrial Economics*, 1981, 29 (3) 247-67.

Goldsmith, E. "Supersedence: Its Mythology and Legitimization." *Ecologist*. Sept./Oct., 1981.

Greene, Brian. *The Elegant Universe*. New York: Vintage, 2000.

Gregory, K.L. Native View Paradigms: Multiple Cultures and Culture Conflicts in Organizations. *Administration Science Quarterly*. September, 1983 359-76.

Guildford, J.P. "Traits of Creativity." *In Creativity and Its Cultivation*, H. Anderson ed. New York: Harper, 1959.

Guthrie, W.K.C. *A History of Greek Philosophy*. Cambridge, England: Cambridge University Press, 1969.

Gawain, Shakti. *Creative Visualization*. New York: Bantam Books, 1985.

Hail, W.K. "Strategic Planning, Product Innovation and the Theory of the Firm". *Journal of Business Policy*, 1973 3 3.

Havelock, R.G. "Planning for Innovation." Center for Research on Utilization of Scientific Knowledge. Ann Arbor, MI: University of Michigan 1970.

Hawking, Stephen. *A Brief History of Time*. New York: Bantam Books, 1988.

Hawkins, Gerald .S. *Stonehenge Decoded*. New York: Doubleday, 1964.

Heisenberg, Werner. *Physics & Beyond*. New York: Harper & Row, 1971.

Herzog, A.R. *Subjective Well-Being Among Different Age Groups*. Ann Arbor, MI: Institute of Social Research, 1986.

Hirschman, E.C. "Symbolism and Technology as Sources for the Generation of Innovations." In *Advances in Consumer Research*. Andrew Mitchell ed. St. Louis: Association for Consumer Research, 1981 9 537-541.

Hofstadter, D.R. *Godel, Escher, Bach: An Eternal Golden Braid.* New York: Vintage, 1980.

—— *Metamagical Themas: Questing for the Essence of Mind and Pattern.* New York: Bantam Books, 1986.

Hofstede, Geert. *Culture's Consequences.* Beverly Hills, CA: Sage, 1980.

Jung, Carl G. *Modern Man in Search of a Soul.* London: Kegan Paul Trench Trubner: 1933.

——.*Man and His Symbols.* New York: Dell, 1964.

Keepin, Will. "Lifework of David Bohm." www.vision.netau/~apaterson/science/david_bohm.htm.

Keirsey, David and Marilyn Bates. *Please Understand Me.* Del Mar, CA: Prometheus Nemesis Books, 1992.

Kuhn, T. *The Structure of Scientific Revolutions.* Chicago: Chicago University Press, 1970.

Langham, Derald G. Genesa: *An Attempt to Develop a Conceptual Model to Synthesize, Synchronize, and Vitalize Man's Interpretation of Universal Phenomena.* Fallbrook, CA: Aero Publishers, 1969.

—— "Genesa." KPFK Radio Interview, 1975.

Laszlo, Ervin. *The System View of the World.* New York: Braziller, 1972.

Lerner, E J. *The Big Bang Never Happened.* New York: Vintage, 1991.

LeShan, L. *The Mystic, the Median and the Physicist.* New York: Viking, 1974.

Lilenfield, R. *The Rise of Systems Theory: An Ideological Analysis.* New York: John Wiley, 1978.

Lovell, Jim and Jeffrey Kluger. *Lost Moon.* New York: Houghton Mifflin, 1994.

Mainzer, Klaus. *Thinking in Complexity.* Berlin, Germany: Springer-Verlag, 1994.

Major British Poets of the Romantic Period. William Heath, ed. New York: Macmillan, 1973.

Maslow, Abraham. *The Farther Reaches of Human Nature. 2nd ed.* New York: Viking Press, 1971.

—— *Towards a Psychology of Being.* New York: Viking Press, 1962.

Miller, J.G. *Living Systems.* New York: McGraw Hill, 1978.

Needham, J. *Science and Civilization in China.* Cambridge, England. Cambridge University Press, 1956.

O'Connor J.S and Ian McDermott. *The Art of Systems Thinking.* San Francisco, CA: Thorson's Publishing, 1997.

Pagels, Heinz. *The Cosmic Code*. New York: Simon & Schuster, 1982.

Peale, Norman Vincent. *The Power of Positive Thinking*. Englewood Cliffs, NJ: Prentice-Hall, Inc., 1952.

Peat, F. David and John P. Briggs. *Looking Glass Universe*. New York: Simon & Schuster, 1984.

Peters, Tom. *In Search of Excellence*. New York: Warner Books, 1984.

Plank, M. *The Philosophy of Physics*. New York: Norton Press, 1982.

Pribram, Karl H. *The Holographic Principle*. N.p.: Freeman Press, 1979.

Prigogine, Ilya. *From Being to Becoming: Time and Complexity in The Physical Sciences*. San Francisco: W.H. Freeman and Co, 1980.

Prigogine, Ilya and I. Stengers. *Order Out of Chaos*. London: Heinman, 1984.

Reich, W. *An Introduction to Orgomy*. New York: Farrar, Straus, and Cudahy, 1960.

Rothman, R.A. *Working: Sociological Perspectives*. New Jersey: Prentice-Hall,1987.

Russell, P. *The Global Brain*. New York: St. Martins Press, 1983.

Sheldrake, Rupert. *Seven Experiments That Could Change the World*. Los Angeles: J.P. Tarcher, 1995.

——.*A New Science of Life: Formative Causation*. Los Angeles: J.P. Tarcher, 1982.

Siegal, Bernie, M.D. *Love, Medicine and Miracles*. New York: Harper Collins, 1990.

Surowiecki, James. *The Wisdom of Crowds: Why the Many are Smarter Than the Few and How Collective Wisdom Shapes Business, Economies, Societies, and Nations*. New York: Doubleday, 2004.

Tomkins, P. *Mysteries of the Great Pyramid*. New York: Harper & Row, 1978.

——.*Mysteries of the Mexican Pyramid*. New York: Harper & Row, 1984.

Tracy, L. *The Living Organization—Systems of Behavior*. New York: Praeger Publishers, 1989.

Von Bertalanffy, Ludwig. *General Systems Theory*. New York: Braziller, 1968.

Walonick, David S. "*A Holographic View of Reality*." www.survey-software-solutions.com/walonick/reality.htm.

____. "General Systems Theory." www.survey-software-solutions.com/walonick/systems-theory.htm.

Watzlawick, Paul. *The Language of Change: Elements of Therapeutic Communication*. New York: Basic Books, 1978.

Webster's New World Dictionary of the American Language. 2nd College Edition. David B. Guralnik, ed.. New York and Cleveland: The World Publishing Company. 1970.

Weiner, Norbert. *Cybernetics.* New York: Wiley, 1948.

Weiner, Philip. *Dictionary of The History of Ideas.* New York: Charles Scribner's Sons, 1968.

Wilbur, K. *The Altman Project.* Wheaton, Illinois: Theosophical Publishing, 1980.

Wolf, Fred Alan. *The Dreaming Universe.* New York: Touchstone, 1995.

—— *Star Wave.* New York: Macmillan Press, 1984.

—— *Parallel Universes.* New York: Touchstone, 1990.

Zukav, G. *The Dancing Wu-Li Masters.* New York: Bantam Books, 1979.